Security of Premises:
a manual for managers

Security of Premises:
a manual for managers

Stanley L. Lyons FCIBSE

Butterworths
London   Boston   Durban   Singapore   Sydney   Toronto   Wellington

All rights reserved. No part of this publication may be reproduced
or transmitted in any form or by any means, including
photocopying and recording, without the written permission of
the copyright holder, application for which should be addressed to
the Publishers. Such written permission must also be obtained
before any part of this publication is stored in a retrieval system of
any nature.

This book is sold subject to the Standard Conditions of Sale of
Net Books and may not be re-sold in the UK below the net price
given by the Publishers in their current price list.

First published 1988

© **Stanley L. Lyons, 1988**

British Library Cataloguing in Publication Data

Lyons, Stanley L.
  Security of premises: a manual for managers.
  1. Buildings—Security measures
  I. Title
  690   TH9705

ISBN 0-408-01367-2

Library of Congress Cataloging in Publication Data

Lyons, Stanley L. (Stanley Lewis)
  Security of premises.

Bibliography: p.
Includes index
1. Buildings—Security measures. 2. Buildings—
Protection. 3. Burglary protection. I. Title
TH9705.L96   1987   658.4'7   87-27825

ISBN 0-408-01367-2

Photoset by Butterworths Litho Preparation Department
Printed in Great Britain at the University Press, Cambridge

Preface

My interest in the subject of the security of premises originated through my activities in developing security-lighting techniques. These commenced in 1968 when, as part of my duties as an industrial technical specialist at The Electricity Council, I started to investigate the crime-prevention value of exterior lighting systems. Soon, I was lecturing on Crime Prevention courses at the Home Office Crime Prevention Centre – a pleasurable duty which continued regularly for over ten years.

Through my work at The Electricity Council, I was privileged to be allowed to mount some large-scale experiments in security lighting techniques, and these soon convinced me that security lighting – though of great value – can only be fully effective as part of a security system which includes the fencing and patrolling of the premises. I attended a number of seminars on crime prevention organised by the Home Office and by local crime prevention groups, and sat in on a number of other people's lectures at seminars on crime-prevention, including liaison meetings between architects, local authorities and the police. As my interest and understanding of security matters grew, I came to see that perimeter defence systems are themselves only part of the solution to the problem of controlling entry. One has also to be aware of management objectives, cost constraints, and – very importantly – the unique features of the particular place and contents to be defended.

By the mid-1970s, I was starting to see 'control of entry' as an entity, a subject in its own right. I read all I could get hold of on subjects such as building construction, detection systems, management aspects of security, and the interaction between the building services and security. It came as something of a surprise to hear myself introduced at a seminar as 'a security expert' – a status I have never claimed, and one which I would be reluctant to accept. There are so many facets to this important subject, so many matters of technology, design and management, that it would be a brave person who could claim to be expert on all. With this disclaimer, I offer the reader this broad summary of my reading and consultative experience in the management of building security.

It was a happy coincidence that the Security Engineering Group of the Chartered Institution of Building Services Engineers formed a Task Group to prepare the section of the CIBSE Guide dealing with security engineering[1] about the time this book was in preparation. I am most grateful to the Institution for making the draft Guide section available to me, for it has influenced my views on some important matters.

Also concurrent with the preparation of this book has been the work of a committee of the British Standards Institution which has been preparing the first British Standards on the security of premises[2], and there have been some valuable exchanges of information.

I gratefully acknowledge the help I have received from the Home Office Crime Prevention Centre at Stafford, where the Director and his staff have been such useful sources of reference and have been most enlightening in discussions of crime-prevention principles. In the preparation of this book I have studied the catalogues and trade literature of a great number of companies who sell security equipment, systems and services.

My acknowledgements go the editors of *Security & Protection, Security Gazette* and *Home Security* magazines, all of whom have published articles of mine on security subjects during the writing of this book. The researches for the book often covered the same ground as that needed for the articles, but I have extended the depth of the treatment of the topics in the present pages. None-the-less, it was most valuable to have earlier exposure of some of the ideas now expounded more fully in this book, and to have had the benefit of the feedback which came from users and practitioners.

On many of the consultancy projects with which I have been involved, other consultants and the Crime Prevention Officers of police forces have made important contributions to the work. Some workers in this field put great emphasis on the uses of electronics in creating security, but I have pursued a path of low technology, trying to devise commonsense methods using the basic ingredients of good engineering to provide physical barriers, coupled with the use of lighting and proper guarding as the main features of defence. I came to see the problems of security as being as much those of organisation and management as of technology. But, working with others on practical projects, a process of exchange of ideas took place, resulting in a growth of knowledge. After some years of work in this field, it occurred to me that the time had come to try to put down on paper a summary of what I had learnt, together with as logical an analysis of the problems and their potential solutions as I could devise, and this book is the result.

While courtesy and gratitude require that I should mention the sources from which I have obtained valuable information, I must make it clear that the views expressed in this book are mine alone. They are, naturally, strongly influenced by my experience as a consultant, first in the field of security lighting, and later working on the wider aspects of security management of premises.

Stanley L. Lyons

Contents

1 Objectives for the security of premises 1
1-1 Practical objectives 1
1-2 How premises may fail in their security function 5
1-3 Analysis of methods of entry 7
1-4 Add-on security hardware 8
1-5 Economics of security 10
1-6 Risk 11
1-7 Who should be consulted in setting the security objectives? 13
1-8 Checklist for review of security 16

2 The defended site area and its perimeter 19
2-1 The approaches to the premises – the surveyed zone 19
2-2 Fencing and walling enclosing the land 20
2-3 Zoning, citadels and redoubts 26
2-4 Exterior security lighting 30
2-5 Alarm systems 33
2-6 Providing facilities for patrolling and surveillance 37
2-7 Relationships with adjacent properties 38
2-8 Control of vehicles; parking 39
2-9 Illegal entry facilitated by topography, weather and seasons 41

3 The building 43
3-1 Functions of the structural enclosure 43
3-2 Access and exit – conflict between requirements 44
3-3 The structural shell and cladding 45
3-4 The floors and foundations 47
3-5 Doors 47
3-6 Windows 48

viii   Contents

3-7   Interior walls and partitioning; ceiling voids   53
3-8   Bridges, walkways, tunnels and external facilities   54
3-9   Electrical services, power supplies and standby power   55
3-10 Interior lighting, pilot lighting, emergency lighting   56

4 Supervision and control of entry   59

4-1   Setting the objectives for the system   59
4-2   Practical difficulties of operation   59
4-3   Checking vehicles   61
4-4   Checking of visitors – preventing walk-in crime   62
4-5   Checking of staff; personal searches, vehicle searches   67
4-6   Response to an alarm situation   69
4-7   Day and night patrolling, internal and external   70
4-8   Use of cctv and radiocommunications systems   74
4-9   Protection of the security personnel   76
4-10 Keys and locks   77
4-11 Control of movement within a building by zoning   80

5 Protection of newly-built and reconstructed premises   82

5-1   Confidentiality at the design and planning stage   82
5-2   Precautions when altering existing premises   83
5-3   Security during construction of new or extended premises   83
5-4   Fencing the site   84
5-5   Power supplies, security lighting & emergency lighting   85
5-6   Security containers   87
5-7   Countering the risks to machinery, cranes and plant   89
5-8   Preventing preparations for future crimes   90
5-9   Security during the early days of occupancy   90

6 Security strategies for typical premises   92

6-1   Offices   92
6-2   Warehouses, storage buildings   93
6-3   Retail premises   95
6-4   Hospitals   96
6-5   Hotels   102
6-6   Security in outdoor environments   106
6-7   Security at premises with especial risks   108
6-8   Public buildings, places of entertainment; sports stadia   111

7 The management of building security 113

7-1 Who is responsible for what, and to whom? 113
7-2 Training the managers 116
7-3 Monitoring security performance 118
7-4 Liaison with the police and security contractors 121
7-5 Liaison with the insurers 122
7-6 Management actions during an ongoing incident 123
7-7 Reports and statements 123
7-8 Planning security management 125
7-9 Confidentiality 125
7-10 Trust 128

References and further reading 129

Index 131

Chapter 1
Objectives for the security of premises

1-1 Practical objectives

This book is concerned with means of preventing unauthorised entry of persons into premises, and the prevention of loss due to all kinds of crime associated with uncontrolled access. Even if cost considerations could be entirely ignored, it would not be practicable to set the objective of totally excluding all intruders, for any system of physical protection might be overcome by main force (e.g. by the use of weapons or explosives, or by threatened or actual violence against the defenders), and even the best security arrangements could be subverted by a breach of trust by those whose duty it is to protect the premises and its contents. Thus, although the objective of total prevention of intrusion is desirable, the practical main objectives must be:

- to make unauthorised entry difficult, time-consuming and hazardous to the attacker; and, where possible, to arrange that he is exposed to risk of detection during all phases of carrying out his crime or attempted crime (and possibly even while making his preparations to enter);
- on the assumption that complete defence of all the valuable contents is not practicable, to accord a higher degree of protection to those assets which are the most valuable or the most difficult to replace;
- on the assumption that attacks will occur, to provide means of surveillance (including the use of automatic detection and alarm devices where appropriate) to ensure a high probability that any attempted attack will be detected early and may therefore be frustrated in whole or in part, with some probability that the criminals may be apprehended, or that they may later be identified, these circumstances being likely to deter those who would attempt breaches of the security of the premises;
- in doing all these things to give highest priority to the protection of the occupants (both the security personnel and all the other occupants), to minimise the risk of their being threatened, attacked, taken hostage etc.

1-1-1 Degrees of protection

Differing degrees of protection may be achieved, according to the capital and operating costs that can be afforded by the occupier or can be justified by the probability of preventing loss. In general, the greater the budget that is available, the higher the level of security that can be purchased; but, at any level of physical protection, and at any level of cost, it must never be assumed that premises can be made impregnable. If the intending intruder has sufficient resources (e.g. know-how, tools, inside information or assistance, courage and time), then it must be possible for the security to be breached.

It may be assumed that the greater the prospect of illegal gain or other satisfaction by the intruder, the greater will be the resources he may attempt to apply to overcoming the defences. If we cannot set the objective of complete exclusion of intruders at all times as the criterion for our success, we can at least define the point where the defences fail – namely, when the intruder not only gains illegal access, but also when he threatens or causes actual harm to the occupants, or damages or steals the assets being defended. If an intruder gains access to the premises, but is frustrated in his objectives of serious damage or theft, and if he causes no injury to the occupants, then it could be said that the security measures have not failed in their most important objectives.

The apprehension of criminals should not be regarded as a prime objective of the security of civil premises. If, as a result of security measures and procedures instituted by the occupier, criminals are apprehended, or if their descriptions are noted for future use in identifying them, or if individuals are identified, then this is a bonus (see 1-1-4). Citizens in the UK and many other countries have the power to make a 'citizen's arrest'; but, in formulating security strategies for premises, the apprehension of criminals should be regarded as being a matter mainly for the police, and requiring the collaboration and assistance of the occupier only as far as may reasonably be expected of him. If the occupier's security measures and procedures assist in this, then so much the better, but their purpose is not law-enforcement; their justification is economic and to protect the occupiers from injury.

1-1-2 Defining entry

In this book, in the use of the term 'control of entry' we include not only incidents where the criminal enters the defended area, but also control of situations where the criminal steals without entering, or where the actual intrusion is limited to passing into the premises something that aids theft or causes damage. Thus we include situations where, for example, the criminal takes something through a window or other opening, or does damage without entering – for example, he introduces a bomb or incendiary device.

We also include situations where persons are normally permitted to enter, but where it is desired to limit their access to certain parts of the premises, or to control their conduct to prevent undesirable acts. Examples of these include airports, where persons may enter the public areas, but they are not permitted access to the technical installations or aircraft in an unsupervised manner, and regulation of movement must be exercised in respect of customs barriers and passport control (see 6-6-3).

At a number of industrial and other sites in the UK, a public footpath or right-of-way passes through privately-owned or occupied premises. Although the occupier has the right to defend his premises against intrusion, it would be illegal for him to prevent a member of the public passing along the footpath. In such cases, the strategy must be to supervise persons passing along the public route, and to restrain them from straying from the authorised route into the defended premises. This control has to be exercised within careful constraints to comply with the laws on trespass.

In the UK, except when committed in defiance of an injunction or court order, the act of trespass is not normally regarded as a misdemeanour or criminal offence, and is not usually prosecutable in the absence of actual damage being caused by the trespasser unless it can be shown that the trespass was committed with the intention to steal or do damage etc. However, the offences of trespass are currently being amended by a new Public Order Act which will take account of 'mass trespass'.

It should be noted that the term 'premises' is used to include land and open spaces as well as walled or roofed areas from which it is desired to exclude unauthorised persons.

1-1-3 Some limitations

In the UK, man-traps are illegal. A man-trap is any device or thing which is intended to trap an intruder, or perhaps also any device or thing which has this effect even if apparently not intended. If an intruder were trapped in a space, or detained on the premises because some part of his person or clothing was trapped so that he could not escape, it seems possible that he could bring a prosecution against the occupier, even if he himself were guilty of burglary or other crimes.

Under the Common Law in England, every occupier of premises has a duty to take steps to ensure the reasonable safety of persons entering his premises – even if they enter without permission or for an illegal purpose. It would be illegal, for example, to wire up electrical connections to a safe to expose a criminal touching it to the risk of electrocution; nor may noxious gases be released; nor may spring-traps, falling weights, nor any other kind of booby-trap or mantrap be employed, such being offences under the Occupiers' Liability Act[22]. It seems likely that the use of such things as barbed tape, anti-scaling barriers and even anti-climb paint could also fall into this category. It would be prudent to place warning notices of the use of such defensive measures.

Although very loud alarm sirens, hooters and bells are used to signal an intrusion and to call assistance, it is not known if it would be possible for an intruder to bring a civil action against the occupier on the claim that his hearing had been damaged by the sound of the alarm. At the time of writing, it appears that no specific upper sound intensity (decibels) is defined as the limit to which any person may legally be subjected, although prosecutions under the Health & Safety at Work Etc Act are instituted against industrial companies on the grounds of their subjecting persons to excessive sound intensity.

Although no positive guidance can be given, it seems likely that the occupier would be safe from prosecution (either criminal, under the Health & Safety at Work Etc Act (HASAWA), or under a civil claim for damages) if he instituted a lock system which had the effect of trapping an intruder in a room in the premises from which he could not escape, provided the intruder was not subjected to any harm, and if he was handed over to the police with minimum delay once his predicament was known to the occupier.

For some situations, the use of guard-dogs is very helpful. A sole guardian, working with a dog, is likely to be more active and alert with a dog than without, and the sharp hearing and acute sense of smell of the dog can be valuable assets. However, in the UK, the deployment of loose guard-dogs on enclosed premises without a handler being in attendance is illegal. The prohibition arises from cases where innocent persons have been injured by such guard-dogs, the classic case being that of a child that was savaged on putting his hand under a gate to try and stroke the dog.

A further curious legal position may also exist in regard to the duty of the occupier to take steps to ensure the reasonable safety of persons entering his premises (*vide* Common Law and Health & Safety at Work Etc Act) in the matter of the safety of security staff. It would require a test case to prove the point, but it seems likely that if a security guard were injured in the course of his duties on being assaulted by an intruder, and if he could show that this occurred because he had not been provided with facilities to ensure his reasonable safety (i.e. security lighting to protect him from surprise attack, a personal radio upon which to call for help, etc), then an injured security guard might have grounds for a claim for damages against the employer or occupier, and conceivably the occupier could also be liable to prosecution under HASAWA.

1-1-4 Typical objectives of occupiers

An informal survey conducted by the author amongst a limited number of occupiers of industrial premises revealed that none of those questioned were the slightest bit interested in catching intruders. What they wanted above all was (a) to deter possible intruders by signs of preparedness, and (b) to scare off intruders by the sound of the alarm system when tripped (and, in some cases, by lights put on for the same purpose).

Of those questioned, over half were more worried about the possibility of an intruder starting a fire than about the risk of loss from theft. The standing orders to the chief of security at one large works specifically instructed him to give low priority to apprehending any outsiders who might get into the works, but to try and identify any of the company's staff or ex-staff who might be involved in crime. He was also instructed to give greatest attention to ensuring that intrusion did not lead to fire.

A fire might be started out of malice, or through carelessness, i.e. from an intruder smoking in a high-fire-risk area – an intruder having such little regard for the premises that he would be unlikely to respect 'No Smoking' notices. A case was cited in which an intruder who was probably smoking when he opened a varnish-stoving oven – possibly merely looking for something to steal – caused an explosion and fire (the intruder was killed). One managing director said that he regarded the perimeter fencing and security guarding simply as ways of reducing the fire-risk at the works – he was of the opinion that a break-in into those particular premises could not be profitable to an intruder, and therefore would most likely be carried out for malicious motives.

1-2 How premises may fail in their security functions

1-2-1 If the defences of the building are breached, or the intruder places occupants under threat or causes them physical harm; or if the intruder damages the premises or its contents; or if he steals something of value, then there has been a failure of the primary functions of security in that (a) the intruder has not been deterred by the signs of preparedness of the occupier to resist attack, and (b) has not been physically prevented from achieving his act of illegal entry.

Failures of security are not always obvious; for example, shrinkage of stock and shortages of cash may be wrongly attributed to dishonesty of the staff or permitted visitors when, in fact, they are due to repeated undetected burglary. Cases have occurred where the same premises have been repeatedly entered over a long period by a former employee who was improperly in posession of keys, the intruder stealing small amounts of goods or money and re-securing the premises, so that the fact that entry had occurred was not obvious (see 4-10).

Penetration of defences which leads only to loss of secret information may never be detected, although the loss of secrecy may disastrously affect the prosperity of the business. Intrusion which leads to the destruction of irreplaceable records (computer records or hard copies) can be very damaging to an organisation, though the intrinsic value of what is stolen or destroyed may be very small (see 6-1-2).

1-2-2 Intrusion into premises may have political or commercial repercussions far outweighing the importance of what is stolen or damaged by the intruder. An example of political repercussion is the case of the intruder who gained access to the bedroom of HM The Queen in Buckingham Palace in 1982, leading to the premature retirement of a senior police officer and considerable embarrassment to the Government.

An example of commercial repercussion is the case of the bank whose vaults were entered by tunnelling from a nearby sewer, the thieves stealing a very high value of property from the safe deposit boxes provided by the bank for its customers, and leading to loss of confidence in the ability of the bank to provide the required level of security. In some situations, public alarm may be occasioned, as when there is even the most minor intrusion into any premises associated with nuclear power or military nuclear weapons.

An example of an intrusion leading to the occupier being the subject of commercial loss and public ridicule damaging to his business is the case of the security company whose premises were audaciously and successfully raided, the wages for the company's own staff being stolen.

1-2-3 Management may be unaware of serious weaknesses in the security of their premises, for few high-status executives will actually get involved; few, for example, will take the trouble to put on gum-boots, and go out and inspect the fence-lines at night. Indeed, in most organisations, it is a rare event for any member of the senior staff to be on the premises during the night hours except in a situation of dire emergency.

6   Objectives for the security of premises

Management's view of security matters is generally a paper one, with all the information reaching the highest levels being first filtered through reports of staff (some members of which might have good reasons to slant the facts or not to tell the whole truth lest they themselves become the subject of criticism).

Well-established 'fiddles' may exist for years, with goods being pilfered, cash being purloined, or employees being off-site while still clocked-on, as well as other malpractices such as unauthorised persons being admitted to the site. As an example of the latter, at a paper mill in Kent the management were quite unaware that prostitutes frequently entered their premises during the night-shift until this was reported by a security consultant.

Another example of unreported failure of security concerned an automobile factory in the Midlands, where careful inspection of the whole perimeter fence revealed no less than seven 'rabbit runs', i.e. unofficial paths leading from the buildings in the complex to weak points in the perimeter fence. Some of the fence breaches were cunningly concealed, and had been used as unofficial routes into and out of the plant for years. Watch was kept, and after three nights of observation, twenty-two persons were arrested and were charged with various offences including theft, being illegally on the premises, or for fraudulently absenting themselves from work while clocked-on. At an in-company enquiry, it was later revealed that it was not part of the company's standing orders to the security staff to patrol along the security fence, nor even to inspect it from time to time.

1-2-4 It is believed that the following circumstances contributed significantly to the decision of a vehicle tyre-maker to close a large factory in the North of England. As part of an investigation into persistent and unexplained low profitability of the plant, all materials entering the factory were weighed. All waste materials were also weighed. By calculation it was possible to make a fairly accurate estimate of the theoretical weight of the products produced. After making adjustments for the weight of packing materials, it was suspected that there was a serious discrepancy in the total weights of goods in and goods out, suggesting that part of the output of tyres was being 'milked' by some kind of fraud or thieving by the staff.

It was decided to keep night observation as part of the preparation for improving the fencing and installing a new system of security lighting. As a result it was soon found that several members of the staff (including a senior manager) were involved in stealing finished tyres by simply wheeling them to the perimeter fence during the hours of darkness, the tyres being lifted over the fence with a block-and-tackle attached to a tree branch that conveniently overhung the perimeter fence. The police apprehended four men in the process of removing stolen tyres, the load on the lorry outside the fence being of several tonnes. Here, the failure of security was both at the physical and management levels, and the enormous losses which eventually led to the closure of the plant were completely preventable.

1-2-5 Concurrent with the writing of this book, women supporters of the Campaign for Nuclear Disarmament (CND) have been continuing a

Objectives for the security of premises   7

long-standing activity at airbases in the UK. The perimeter fences at such premises are commonly standard BS 1722: Part 10 fences, 2.6 m high and topped by three strands of barbed wire[3]. Such fences have been repeatedly breached by protesters who have either been removed from the site or arrested. The fence forms no more than a convenient demarcation between the defended area and the outside world, and its effectiveness depends entirely on the continuous presence of security guards and police. Technically, the fence 'fails', in that the protesters are not prevented from entering. However, the system as a whole (at the date of writing) has not failed, because of the defence-in-depth due to the provision of inner fence lines (see 2-2), security lighting (see 2-4), citadels (see 2-3) and detection devices (see 2-5), and because of the vigilance of the security patrols and police supervision (see 2-6, 4-7).

1-3 Analysis of methods of entry

1-3-1 In this section, we look at the generic ways in which uncontrolled entry into defended premises might be effected. Specific methods of entry include dangerous and difficult routes (e.g. leaping at roof level from one building to another, entering via sewers or by tunnelling etc) which are likely to be used only in situations where there is such a highly-prized target that the risks seem justified to the intruder, or where the motivation of the intruder is fuelled by extreme passions and not merely a simple lust for gain. Before attempting an analysis of possible entry routes, it is constructive to classify the operational method of the intruder:

- Entry may be attempted or effected by overt force or the threat of force, e.g. by threat or use of firearms.
- Entry may be attempted by trick, e.g. by presentation of false documentation or by 'passing-off'. Passing-off means simply entering and looking as though one belongs, perhaps carrying some official-looking prop and muttering 'Electricity Board!' or some other such inferred authority. Uniforms and overalls, such as those worn by various kinds of legitimate visitors, are easily obtained and are very effective if the security staff are not well trained and vigilant.
- Entry may be surreptitious, and may simply involve walking into undefended premises, or may involve an act of breaking and entering. Surreptious entry is more likely to occur at times when the building is not fully staffed, e.g. at night, or when the attention of the protecting staff is diverted. Criminals may create diversions to make the opportunity for entry (e.g. they may start a fire in some rubbish nearby, or an accomplice may pretend to be having a fit, etc) to capture the attention of the guard for long enough to steal keys, or to slip into the building.
- Entry may be made legally as a preamble to committing a crime. For example, the criminal may legally enter the public access area of an office building or other premises, and then force an entry from that area into a locked room, or commit any kind of crime while still in the public-access area.

An intruder who has effected a surreptitious entry may attempt to bluff his way out if challenged, or may make an attempt to use force. In a

8   Objectives for the security of premises

complex and well-planned crime, intruders may be operating in one or more of these modes at one time, and may switch from mode to mode during the course of the crime.

Scenarios may include entering by force, and then mingling inconspicuously with a crowd until the time for the next stage of the crime; or entering legally, e.g. entering an airport with a valid ticket, and then carrying out an act of violence. In the latter case, if the credentials for entry are easily obtained (as is an airline ticket), then the entry cannot be prevented; but the security staff will need to control the actions of persons who, having gained legal entry, then proceed to perform a crime. In such cases a prime security objective will be to control the movement of persons between zones within the defended area (see 2-3).

1-3-2 Analysis of the methods of entry can be furthered by consideration of the point of physical intrusion into the defended area, thus:

- Entry may be made through a normal point of entrance and egress, e.g. works gate, office main door etc.
- Entry may be made by a route that is normally designated as an emergency exit route, e.g. a fire exit door.
- Entry may be made through an unusual or temporary route which provides an opportunity not normally available, e.g. via scaffolding provided for repairing or cleaning the building, or through an opening in the perimeter fence or structural wall made for the purposes of modification or repair.
- A forced entry may be made involving damage to the premises, or without damage by use of keys or picklocks etc. Such entries may be made through the perimeter defences into the defended zone, or entry may be made from a defended zone or from a normal public-access area into a citadel, to attack a target within that or another zone.
- Entry may be made at a time and under such conditions that the entry is legal, and the criminal may then remain on the premises concealed or by trick for illegal purposes.

(For review of the defensive functions of the perimeter fence, see Chapter 2; for review of the defensive functions of the building structure, see Chapter 3. See also Trust, section 7-10.)

1-4 Add-on security hardware

1-4-1 In the case of new premises or those being structurally modified, pressing matters such as the need to complete the project quickly, or to restrict capital expenditure, may lead to the deferment of the important stages of planning the security strategy. Instead of devising a proper security plan, the occupier may purchase some add-on security hardware which he hopes will give him immediate and effective protection. It may work; but the best course will always be to think the whole security problem through, to study the nature and magnitude of the risks, and to attend to any structural weakness which would make the property vulnerable to attack. Then, as required, additional items, such as alarm

Objectives for the security of premises 9

systems, security lighting etc., may be added on a scale commensurate with the resources available and the degree of risk.

It is the common experience of Crime Prevention Officers that they often are called in to advise on protection of premises only after the first break-in; but, that first break-in could be disastrous to an organisation. It is more sensible to incorporate such structural security features as will thwart or discourage criminal attacks, or which will minimise losses, even if these add to the construction time and the initial cost of the building.

It is tempting to add-on new security gadgets one by one to overcome real or suspected weaknesses in the defences. Rather than buying additional hardware which the salesman assures you will counter a threat, a more logical course may be to eliminate the risk entirely by hardening the target in some way. It is always better to keep the thieves out by structural and tactical means, rather than to employ a device which rings a bell to tell you that they have got in.

1-4-2 Those responsible for security are constantly assailed by advertisements for systems and hardware claimed to upgrade security. Skilled salesmen who purvey security systems and equipment know how to present their products so as to tempt the buyer to make an instant decision to buy rather than to carry out an in-depth study of his needs. This does not mean that very rapid improvement – that day if necessary – should not be taken to counter a newly-discovered risk.

As an example of prompt response to a new situation, consider the case of the petroleum storage depot which was attacked by political extremists one night. The damage was slight but, unfortunately, next day a newspaper carried a story about the incident, including an ill-advised interview with a security guard who stated, 'It's as black as pitch here at night, you know. The fences are rotten – anyone can get in. If they had attacked that tank there – not this one – it would all have gone up with a bang, I can tell you!' With the vulnerability of installation so publicised, immediate action was necessary. This took the form of moving in extra security staff, and procuring several trailer-lights (see 2-4) to illuminate the vulnerable area temporarily at night until permanent security lighting and new fencing could be installed.

1-4-3 If a risk occurs only infrequently, then it could be sound strategy and good economics to buy or hire some portable or temporary means of countering the threat at the times of special risk rather than undertake structural changes to the building to improve its defences against attack. For example, a temporary exhibition of valuable art treasures, held in a building that is not normally used to house items of high value, might be protected with portable intruder-detection equipment and by the provision of extra manning for the vulnerable period. For such applications, acoustic or passive-infra-red equipment could be installed without the need to make a permanent installation. Also, where the situation warrants it, it would be possible to install some temporary exterior security lighting to give protection to the approaches to the defended premises. But, if it were decided to hold art exhibitions regularly in the building, it might be far more effective and cheaper in the long term to harden the target by beefing-up the physical defences, fitting stronger doors and locks, and reviewing all possible routes of entry into the building, as well as providing such lighting and intruder-detection systems as may be appropriate.

1-5 Economics of security

1-5-1 Doubling the expenditure on physical measures of security and guarding will not necessarily halve the risk of intrusion, though insufficient funding will make it impossible to achieve a level of protection appropriate to the risk. The objective should be to allocate sufficient cash for initial capital expenditure and to budget for the continuous cost of guarding to enable the premises to be so equipped and manned that effective protection is achieved. Neither the capital outlay nor the planned operational expenses can be regarded as fixed sums; the capital equipment will require repair or improvement from time to time, and eventually may need to be replaced in part or in whole; further, the operational expenses will probably increase in line with inflation. By study and the application of the concept of discounted cash flow (DCF), it may be possible to make acceptably accurate projections of cost.

1-5-2 Success of such economic planning may be judged by comparing the actual loss sustained in a period with the potential loss (i.e. the loss that almost certainly would have occurred if the expenditure had not been undertaken). The cost calculation should include (a) the capital cost, amortised over the review period, with adjustment for the loss of the interest that the capital sum would have earned (DCF), or the amount of interest to be paid if the capital sum is borrowed; (b) operational expenses (labour, repairs, maintenance, energy cost etc, inflation-adjusted), and (c) cost of insurance, inflation-adjusted. It is this total expenditure, taken over the review period (say, ten years) that must be compared with the potential loss. The term 'capital cost' includes not only discrete items such as security hardware, but also any extra costs due to alternative designs and construction methods incorporated into the building plan to facilitate security objectives.

1-5-3 If the potential loss due to theft or damage was £1M, and the total expenditure over the review period (see 1-5-2) was £100,000, it might be said that the loss-protection factor was £100,000/£1M = 10%. There are two weaknesses in this reasoning: (a) in practice, it is not possible to trade off the cash saved by not providing physical defences against the higher cost of insurance, because failure to provide adequate protection by physical measures and guarding may so inflate the insurance cost as to make it totally unacceptable; (b) if there were no losses in a period, it cannot be assumed from this evidence alone that the security measures are effective and cost-effective. Despite the good record, there could be a major break-in with serious losses on the very night such an assessment was made.

Objectives for the security of premises   11

1-5-4 If you inform your insurers that you have installed improved means of defending your property, they are unlikely to reduce your premiums immediately. However, if the new measures are in fact accompanied by reduced losses or freedom from loss, it is possible that the rate of increase of future premiums will be slower, and conceivably the premiums might actually reduce over a period. The occupier could test the situation from time to time by getting quotations from other insurers.

Following a costly break-in or incident involving loss against which claims are made, or when applying to an insurer as a 'new risk', it is quite likely that the insurer will impose conditions regarding the physical defences and guarding precautions that are to be taken; and a higher premium may be imposed if the insurer's requirements are not met. Further, if security measures are not instituted to the satisfaction of the insurers, the occupier may find that his cover has been reduced by the insurer, or – exceptionally – a claim might be refuted or discounted on the basis that the occupier has not taken reasonable precautions to protect his property.

1-6 Risk

1-6-1 Factors of risk

A threat is the possibility of the occurrence of an adverse event (such as theft, arson, criminal damage, assault etc.) which would result in personal injury or financial loss. Risk is the assessment of the probability of a threatened event happening. Risk management is the art of minimising the losses which may occur.

To appreciate how one might, for example, assess the risk of theft from a break-in, consider any valuable object. Some factors which will enhance the probability of it being stolen include the following (in random order):

- that the object has sufficient value to tempt the thief;
- that it is small enough to be easily transported away, or that means can be found to overcome the disadvantages of size or weight;
- that the object can readily be sold to an unscrupulous buyer;
- that the object is not easy to identify positively if later recovered by the police;
- that the measures of physical protection are weak, and can be readily overcome to effect surreptitious entry, or entry may be easily gained by force, or by trick;
- that any alarm devices can be overcome by skill, or by taking advantage of the failure of the defenders to set these, or by carrying out the theft so speedily as to escape within the response time of the police or defenders to the alarm;
- that persons of dishonest nature know that the object exists, and its location, defences etc;
- that the defensive actions of those set to guard the object may be overcome by trick, by force or the threat of force, or by their negligence;
- that the nature and location of the site of the valuable object is such as to

expose the criminals to little risk of being noticed in their approach, during the crime, or while making their escape;
- that the persons set to defend the valuable object are not trustworthy (see 7-10).

1-6-2 Motivation for crime

To study the risks in a particular situation, it could be constructive to consider what might motivate a criminal. For example, theft may be performed for such reasons as the following (in random order):

- desperate need (e.g. theft of food or necessities);
- personal gain (to keep or sell the goods for profit);
- avarice (e.g. desire to possess the object);
- duress (e.g. made to steal by the threat of another person);
- jealousy (e.g. satisfaction in taking from the owner);
- malice (e.g. to steal or destroy in order to harm the owner).

Some attacks upon buildings, their contents and personnel, do not appear to benefit the perpetrators at all. Some criminal acts, such as arson and criminal damage, are performed without any apparent motivation. Others may be carried out for revenge by aggrieved present or past employees or associates of the company. Break-ins to simulate attack by outsiders may be made to conceal the fact that goods have been stolen by employees. Arson may be committed to conceal stock shortages, cash deficiencies or falsification of records. Fire may be the outcome of a break-in (see 1-1-4).

There may be an unsuspected risk of attack from politically motivated persons, or by persons pursuing some extremist philosophy. Apparently innocuous premises and companies may be subjected to violent attack by persons supporting extremist organisations connected with religious views, support of animal rights, women's rights etc., or even by pickets in industrial disputes. It is impossible to defend fully against the bizarre, motiveless and illogical attacks which may be made by the mentally deranged.

1-6-3 Assessing the risk

The occupier has to make decisions regarding the methods he will employ to protect his staff, the premises and their contents. He has to decide how much to invest in physical security and in manpower to achieve the level of security he considers necessary, and will seek to spend only on cost-effective methods. Failure to provide the protection will result in higher insurance premiums or difficulty in obtaining effective insurance (1-5-3).

When setting up new premises, or carrying out structural changes to existing ones, or making changes within the premises which will affect the risk (e.g. introducing new processes, increasing the value of stock held, installing new equipment), it would be wise to discuss the matter as early as possible with one's insurers. Generally, insurance companies will not give a

Objectives for the security of premises 13

discount if you install good security devices and measures; it is the history of claims that tends to determine the future premium. Some risks which may be acceptable to the insurers may be unacceptable to the occupier; for example, loss of secrets could not be compensated-for by insurance, and the threat of attack upon employees is a risk that the occupier will wish to minimise. It is the assessment of the potential loss (see 1-5) that provides the justification for suitable investment in physical security measures. Mathematical aids to assessing risk may be used[1].

1-7 Who should be consulted in setting the security objectives?

1-7-1 The objectives for the protection of premises against crime may be ill-defined at the outset of the project, and commonly are not agreed between those affected before site work commences. Consequently, the required level of security may not be achieved initially (though improvements may have to be instituted with urgency – perhaps after an early experience of loss – a classic case of locking the stable door after the horse has bolted). If the plan for security is not determined at the outset, the final cost of installing and operating the security system may be far greater.

1-7-2 In designing a new building, the architect is unlikely to devote particular attention to the subject of security unless he is properly briefed. Without clear briefing, the architect may fail to provide the physical strength of shell required to resist penetration, and his selection of window and door equipment may not take account of the security level necessary. It is not uncommon for architects to incorporate design features which enable intruders to climb the building or to find places of concealment that aid crime; or the configuration of the layout may make supervision difficult.

It is quite common for serious discussion of security matters to start only when the outline scheme for the premises is presented by the architect for approval, or – far worse – after the building has been constructed. Things may be rather better in the case of larger organisations which can refer back to the security record of similar premises, and – in the case of premises being refurbished – the security record on the actual premises before the current work was begun. It cannot be emphasised too strongly that study of the security of the building should be made at the very earliest stage of design planning, rather than to try to add security features later (see 1-4).

1-7-3 Consultation on security strategy and specification may take place with the following:

- *The client* The client should outline the functions of the premises, and make an early estimate of the value of the goods and plant which will be at risk. After preliminary discussions with his advisors, the client will also be setting budgets for the whole project and for the cost of the security features. His decisions at this stage will affect both the nature of the security measures, and the cost of the maintenance and manning of the security function during the life of the premises.

- *The architect* Within the general briefing given by the client, the architect will be required to put forward one or more outline schemes for discussion and approval. It is important that the security aspects of his brief should be one of the criteria by which the suitability of his proposals is assessed. It is advisable for the architect to be in touch with the Crime Prevention Officer of the police. Many police forces now employ Architect Liaison Officers to provide architects with this type of information at an early stage of the development of projects.

- *The building services consultant* This may be one firm of multi-discipline consultants, or there may be a number of specialist consultants dealing with such services as heating, ventilation, air-conditioning; electrical distribution; lighting; fire-precautions, escapes, sprinklers, alarms; communications, telephones, computers; architectural details, doors, windows, roofing, partitioning, interior finishes. Ever more commonly these days, the building services consultant may provide or co-ordinate advice and specifications for the security aspects of the structural features as well as the specification for security hardware. Specialist consultants may also be employed (see 1-7-4).

- *The Crime Prevention Officer (CPO) of the local police* The CPO will bring to the discussions a detailed knowledge of the district and its crime record. Security precautions suitable for premises in a quiet country town would probably be quite inadequate for similar premises located in a city centre. Similarly, the level of security required for say, a bakery, would be inadequate for a warehouse containing thief-attractive goods such as video recorders or tobacco. The CPO will be informed on such matters, and also on the latest methods being used by criminals, and will have knowledge of any special local circumstances, e.g. relationship of the site to special risk areas of low-grade housing, or means of entry through derelict land or buildings adjacent to the site.

- *Staff security manager* It is desirable that the person who will operate the security system of the premises initially should be in the consultative team. If a fully-experienced person is to take this post, or if the appointee is someone who has had experience of the organisation's past security record, his advice will be of particular value. He will, for example, be able to provide estimates of the manning levels required to achieve the projected security standard in conjunction with various alternative systems and equipment which may be under discussion.

- *Insurers* In the UK, the major insurers have highly-trained inspectors and assessors, and commonly their advice is available without charge to the insured. They may provide guidelines for the physical security of the premises, but insurers cannot be expected to give recommendations in detail nor to write the security specification. Consultations with the insurers should take place as early as possible in the planning of new buildings or structural changes to existing premises. Note that insurers may accept risks that are unacceptable to the occupier, and conversely; though essentially their objectives are similar, each sees the problem from a different viewpoint.

● *Fire Prevention Officer* There is often a conflict between the requirements for security and those for escape from premises in emergency or fire (see 3-2). The advice of an expert on fire precautions can be of great value.
● *Security consultant* On larger projects it is becoming more common these days for a security consultant to be appointed. Such a person will be qualified and fully experienced, and possibly will be a building services engineer who specialises in security work. Retired police officers often have valuable inside knowledge of how crimes are committed, but may lack the technical expertise required for dealing with modern security installations.

1-7-4 Building services consultants and Crime Prevention Officers should not be expected to provide detailed specifications for specialist equipment such as closed-circuit television (cctv), and it may be necessary to seek the advice of persons skilled and experienced in such areas. Commonly, advice of this kind is sought from potential suppliers, and great care must then be exercised to sift salesmanship from genuinely sound advice. Alternative proposals must be compared, and this may be far from easy if the offers are couched in technical terms and differ in detail one from the other. Even fully-experienced building services consultants should not be expected to be able to provide detailed designs for services such as security lighting installations for large, complex or very high-risk situations. For such projects, specialist consultants may have to be employed. If equipment is offered competitively by several manufacturers, their proposals must be carefully evaluated and, again, the commercial objectives of the potential suppliers must not be forgotten.

1-7-5 Where it is necessary to seek specialist advice from suppliers and contractors, it may be a wise course to arrange to pay a consultancy fee for their advice, with the proviso that part or all of that fee shall be rebated if that company succeeds in becoming the supplier. If this is not done, one arrives at the most unsatisfactory situation where a company which comes on the scene later may be able to underbid another company which has already invested considerable time and effort (and therefore cost) in assisting the client to arrive at a specification for the goods or installation in question. This latter practice (known as 'skinning the contract') is a cause of reluctance on the part of specialist suppliers to provide their best efforts in advising on proposals and designing layouts at the preliminary stages of the project.

An alternative method is to appoint one reputable supplier as the 'nominated supplier', the goods and services to be supplied on a 'cost plus' basis. A further and increasingly popular method is to call for tenders to 'design and build' or 'design and install' the equipment or structures required. In such a case, those submitting tenders must be provided with a specification upon which their offers are to be based. Once the successful contractor or nominated supplier is appointed, any required modifications to the proposals can be introduced at the discretion of the client. This has the hidden advantage that staff of the quoting organisations (other than the successful one) will not have complete knowledge of the final plans.

16 Objectives for the security of premises

1-8 Checklist for review of security

When security matters are reviewed at progress meetings during the planning stages of new projects, or reviewed at periodic surveys of existing premises, thought should be given to the risks associated with each item, and a decision made as to whether the risk is acceptable, or whether something should be done to remove or counter the risk. For example, a vulnerable window could be:

- made stronger by replacing a wooden frame with a metal one;
- fitted with strong locks;
- reglazed with polycarbonate sheet instead of glass;
- fitted with an intruder-detection device;
- fitted with bars externally or a grille internally;
- made more difficult to reach from outside, i.e. by putting barbed wire below it, painting adjacent stackpipes with anti-climbing paint;
- made non-openable, i.e. by filling in the window opening with toughened-glass bricks, or simply bricked up and entirely dispensed with (but seek the advice of the Fire Prevention Officer before doing this).

For each physical feature, decide if the benefit of having that feature outweighs the risk, i.e. is the entry of fresh air and daylight at this point worth the security risk of having a window in such a vulnerable position? And will the cost of installing physical protection measures be justified by the benefits from retaining the window? Consider:

- Is the chosen site for premises suitable for the operation of the proposed activities with good security?
- Is there adequate space around the structures to permit the devising of suitable systems of fencing (including the creation of internal security zones), lighting, alarms and the movement of patrols as may be needed?
- Are there any boundary fences owned by others, and, if so, are they of suitable design, or can they be brought to a suitable standard of security with the cooperation of neighbouring property owners or the local authority? (see 2-7). Or will land have to be sacrificed by setting up additional fences inside the perimeter fenced by others?
- Are there any covenants on the land, or any planning or other constraints which might prevent or make more costly the achievement of suitable standards of security? For example:

  ○ Because of proximity to a railway line, airfield or roadway, will it be possible to devise a suitable type of security lighting?
  ○ If adjacent to a river or navigable water, can lights be shown to seaward or over the water? Will high tide or low tide conditions facilitate illegal entry?
  ○ Is it permitted to fence along a river bank or sea shore, or is there a right of public access?
  ○ Are there any footpaths or public rights-of-way over the land?

- Will it be possible to exercise supervision over land external to the site, i.e. will the surveyed zone extend beyond the perimeter? Consider:

  ○ Are there any obstructions to visual supervision externally?

Objectives for the security of premises 17

   ○ If there are visual obstructions, are they permanent? If they can be removed, at what cost?
- Will seasonal changes affect the security of the site? Consider:
   ○ Are there any marshes or a river etc which may become frozen in winter and permit access?
   ○ Will foliage obstruct visibility in summer?
- Would collaboration with occupiers of adjacent properties enable the security to be enhanced? (see 2-7). Consider:
   ○ Could arrangements for mutual security routines and surveillance be made with neighbours?
   ○ Is it possible to devise a common system of fencing and security lighting with neighbours?
   ○ Will neighbours object to any outward flow of light onto their property from your security lighting?
   ○ Could the neighbours or the local authority provide improvement to the lighting of adjacent sites or roadways?
- Are existing fences of suitable design, strength and height for the assessed risks? Consider:
   ○ Are existing fences in good repair?
   ○ Has a detailed survey of existing fences – from inside and from outside, by day and by night – been carried out recently?
   ○ Are there any signs of 'rabbit runs', i.e. unofficial paths leading to weak spots in the perimeter which are used by staff, criminals or trespassers?
   ○ Is vegetation or foliage masking the view or growing close to the fence-line? Have trees near the fences (which might facilitate entry or the illegal movement of goods) grown since the last inspection?
   ○ Are there any new activities outside the fence-line, e.g. new roads, new structures being built or completed, new occupiers of adjacent land, which might merit further review of the security situation?
- Entrances to the site. Consider:
   ○ Are all existing entrances necessary?
   ○ Are all site gates capable of being opened or securely closed quickly in an emergency?
   ○ If a 'lock' is provided at any entrance, is the distance between the outer and inner gates sufficient to hold the longest vehicle (plus trailer) that is likely to pass through, with adequate side fences and lighting?
   ○ If there is a fast road outside the gates, is there sufficient room at the gate to enable a driver to pull off the road to a safe spot while his papers are being examined before admission?
   ○ Can pedestrians slip through the entrance while a vehicle is in the 'lock' being examined? Is a separate gated pedestrian entrance route required?
   ○ Can the road surface within the 'lock' be painted with white plastic paint to afford reflection of light to the undersides of vehicles for search purposes?

18   Objectives for the security of premises

- Gate-house, checkpoint hut. Consider:

    o Can the security guard see the whole of the entrance area easily from within the gate-house? Is the lighting within the gate-house sufficient for the guard's purposes without revealing him to an outside observer? Can the windows and lighting be arranged so that the occupants of the gate-house are not visible in silhouette between two windows on opposite sides of the hut?
    o Is there need for one-way-vision screens or protective grilles at the windows?
    o Could the windows be tilted slightly to reduce internal reflections?
    o Are there functions being performed in the gatehouse (e.g. supervising fence-alarm panels) which would be better done in a central security control room away from the perimeter?

- Communications. Consider:

    o Does all the equipment work adequately? Do staff know how to use it?
    o Is an ex-directory line needed in the gate-house?
    o Warning buzzers and telephones – can they be heard outside the checkpoint hut?
    o Cctv monitors, mimic diagrams and alarm indicator panels: are they clearly readable by day and night?
    o Can any cctv monitor, mimic diagram, alarm indicator or glass-fronted key cupboard be seen from outside the hut (and therefore be helpful to a criminal)?
    o Bleepers and two-way personal radios: are they needed? If in use, do staff understand their use and also know their disadvantages?

- Key security. Consider:

    o Are keys always kept in a locked key cupboard?
    o Are the key routines adequate and followed faithfully by all staff?
    o Is the routine for dealing with a lost-key situation understood by all concerned and always instituted immediately it is known a key has been mislaid?

- Monitoring and reporting. Consider:

    o Are proper records being kept up during the progress of all occurrences?
    o Are the records of all security matters and occurrences kept confidential and secure?
    o Are there efficient management routines to ensure that correct actions are initiated in response to reports from the security staff?
    o Is the person in day-to-day control of security of sufficient status and power within the organisation to carry out his duties effectively?
    o Are top management and the boardroom kept up to date with all important matters relating to security?

Chapter 2
The defended site and its perimeter

2-1 The approaches to the premises – the surveyed zone

2-1-1 Although the occupier of premises is responsible only for security within the area encompassed by his own perimeter (which we term the 'defended zone'), in practice he should be keenly interested in all that happens on the approaches to his premises.

Lighting and detection systems will centre on the fence, including a strip of land 2 m wide either side of the fence – this 4m-wide strip being termed the 'fence zone' which an attacker has to enter to penetrate the fence from without or within (see Figure 2.1). The fence zone should always be completely free of vegetation or any object which could give man-cover. Outside the fence zone, a distance of, say, 20 to 50 m deep will be subjected to some measure of surveillance, either by observation from within the defended zone, by observation during perimeter patrols inside the fence, or by patrols outside the fence in the surveyed field if access for this purpose can be obtained (see 2-7).

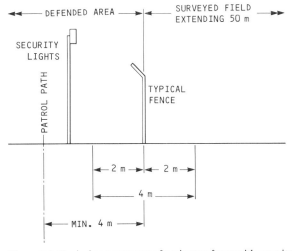

Figure 2.1 Typical arrangement of perimeter fence with security lighting

20 The defended site and its perimeter

2-1-2 Because the keen security guard will consider that all vehicles and other objects near the perimeter of the defended zone might facilitate crime, it should be a matter of routine to observe, record, and be prepared to take action regarding any person or object giving rise to suspicion – even if the premises are not actually under attack. Indeed, it could be wise to assume an attack upon a perimeter fence to be imminent or to have started when an unknown person approaches stealthily and gets within a few metres of the perimeter fence.

Where permission from neighbouring occupiers can be obtained (see 2-7), it is wise to arrange for the security lighting to distribute some light outside the perimeter into the surveyed field, so that persons loitering, suspicious objects and vehicles etc. can be seen. Surveillance may also be carried out by cctv or other detection systems (see 2-5, 4-8).

2-2 Fencing and walling enclosing the land

2-2-1 The degree of delay and impediment offered to an attacker by a perimeter fence depends upon the construction of the fence, and the standard of patrolling and guarding. Ordinary chain-link fences can be cut with a simple pair of wire-cutters; yet such fences will withstand considerable attack (for example, by the actions of aggressive strikers or extremist protesters) by virtue of being supervised by police and guards. Fences may not only be cut and penetrated, they may be scaled (with or without use of apparatus); they may be bridged (e.g. from the branch of a tree, with a ladder, or by use of a crane); they may be pulled up at the base so the intruder can roll under; tunnelled under; or pushed or pulled over by persons, animals or vehicles. None the less, fencing systems enable satisfactory standards of premises security to be attained.

2-2-2 Perimeter walling and fencing should be planned as part of the security strategy on the clear understanding that no perimeter wall or fence is invulnerable to determined attack. The objectives of erecting a fence are:

- to declare clearly the limits of the defended area;
- to make unauthorised entry physically more difficult for the intruder and thus slow down the attack (thereby exposing the attacker longer to the risk of detection);
- to make attempted or actual entry more easily detected by the defenders;
- to make it more difficult for stolen goods to be removed from the premises;
- to make it more difficult for employees and other persons to leave the site other than through the proper controlled channels;
- to create an enhanced possibility of apprehending intruders. Catching an intruder may depend on how well the fence keeps him in once he has penetrated it; if he can quickly cut or otherwise breach the fence, the probability of his escaping after detection is greater.

It is the combination of all the above factors that forms a further less tangible objective:

- to deter the criminal from attempting the attack.

2-2-3 Before going into technical details of fencing systems, it will be as well to consider the defence strategy called 'Fences, lights and men'. The steps of the argument are as follows:

(1) Guards would find it difficult to defend the premises without the aid of the physical obstruction of the fences.
(2) Fences will not keep people out if there are no guards to defend the premises.
(3) Guards can easily be evaded in darkness, so there is a need for security lighting.
(4) security lighting by itself will not keep intruders out.
(5) Thus, it is the combination of 'Fences, lights and men' that can provide an acceptable measure of defence of premises by day and by night. Indeed, because of the glare effects and concealment effects of security lighting (see 2-4), the premises may in fact be more secure at night than by day.

2-2-4 While it might be thought that the objectives of 2-2-2 would be best served by building strong high walls of brick, stone or concrete, it should be realised that solid walls have significant disadvantages to the defenders, including:

- Solid walls prevent the defenders maintaining easy surveillance of the surveyed field outside the perimeter (see 2-1-1).
- Solid walls may be climbed, especially with climbing aids, and the intending intruder will not be seen in his approach. Reaching the top of the wall, the intruder can peek over to see if any guards are present before entering.
- Solid walls tend to muffle the sound of approaching intruders and vehicles.
- The capital cost of solid walls may be six or more times greater than the cost of typical fences of chain-link or palisade types.

The choice of type of fence will be influenced by the site conditions, the level of security desired, as well as budgetary considerations. Types of fences are specified in BS 1722[3].

Three types of 'see-through' fencing are reviewed here: chestnut paling fencing for low-risk and temporary applications (see 2-2-5); chain-link fencing which is commonly-used for ordinary commercial and industrial sites (see 2-2-6); and metal palisade fencing which is used for high-security applications (see 2-2-7). Although not specified in BS 1722, 'weld mesh' fencing is gaining popularity due to its durability, ease of erection and resistance to vandal attack. It is functionally similar to chain-link fencing, but more robust (see Figure 2.2).

Conceivably, a simple and economical fencing method such as chestnut paling might be used to fence the empty site, and be replaced with chain-link fencing later; possibly chain-link fencing might be installed initially, and replaced with more costly metal palisade fencing at a further stage of development of the site. The advantages of 'see-through' fencing are:

- The defenders can keep surveillance on the surveyed zone. Conversely, the villain outside can see in, but this may not be a disadvantage if he sees

22 The defended site and its perimeter

Figure 2.2 An example of 'Expamet Universal' security fencing topped with anti-climbing barrier used as perimeter fencing around a car factory. (Photo: The Expanded Metal Company Ltd, Oakwood, Unit 1, 205 Old Oak Common Lane, London, W3 7DX)

frequent signs of defenders; further, tactical concealment for the defenders can be achieved by means of suitable security lighting (see 2-4).
- By day, and at night too if suitable lighting is installed, the intending intruder can be seen during his attack.
- Open-construction fences do not muffle the sound of persons or vehicles immediately outside the perimeter.

2-2-5 Chestnut paling fencing. If the security risks are not great, and for short-term use (e.g. for enclosing a construction site), it may be possible to employ chestnut paling fencing (also termed 'wooden palisading'). This consists of pointed wooden stakes joined together laterally with lines of twisted galvanised steel wire. The fence is supported by metal or wooden posts at intervals, and its construction can be reinforced with two strong arris rails. Such a fence may be topped with runs of barbed-wire or aggressive barbed tape, and, if properly installed and snugged to firm ground, it gives a reasonable degree of discouragement to intrusion. Chestnut paling has advantages such as:

- Lengths of fencing can be rolled up for storage, and are easily transported to site.
- Can be erected quickly by unskilled labour and at low cost.
- Easily dismantled for further use elsewhere.

The defended site and its perimeter   23

The disadvantages of chestnut pale fencing include:

• Fencing is easily damaged by collision by vehicles.
• It is inherently insecure if erected on soft ground, and it is difficult to make a reasonably secure base unless a wooden rail is fixed to the bottom and staked down to prevent persons – particularly children (see 5-4) – getting under the fence.
• A rubbish fire placed injudiciously near the fence may result in the fence being damaged.

2-2-6 Chain-link fencing. The common type of chain link fence employed at many industrial and commercial premises is that specified in BS 1722 Part 10: 1972[3]. While this type of fence can be scaled or cut, it does provide a level of security that is satisfactory for situations where the risk is moderate, and particularly where security lighting is installed and patrolling is carried out. Chain-link fencing has relatively low initial cost, being about one third of the cost of metal palisade fencing (see 2-2-7), and having low upkeep cost; further, because the groundworks are limited, fences of this kind can be taken down and re-sited to suit changed requirements.

Chain-link fencing will preferably have the mesh material pvc-coated for weather protection. The choice of colour will preferably be dark green or black to reduce the reflection-factor and hence the luminance of the fence at night (and thus facilitate the guards being able to see through it: see 2-4, 2-6). The risk of the mesh being cut is greatly diminished by specifying heavyweight gauge – stout wire up to 3 mm diameter can be obtained. For certain high-security situations a double fence-line may be employed (Figure 2.3). Intruder-detection devices can be applied to chain-link fences (see 2-5).

Standard fencing of this kind, 2.6 m high and topped with three strands of barbed-wire (or barbed aggressive tape in the case of high-risk areas) is

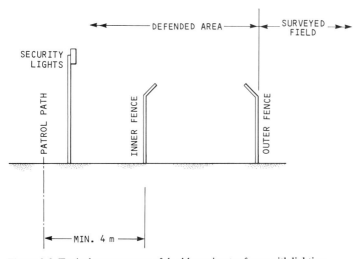

Figure 2.3 Typical arrangement of double perimeter fence with lighting

24   The defended site and its perimeter

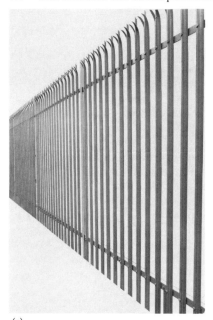

(a)

(b)

Figure 2.4 (a) Section of steel palisade fencing; (b) formation of a citadel of steel palisade fencing to protect a vital point; (c) fantail to protect end of a fence meeting a canal; (d) climbing obstacle formed out of steel palisading to protect a vertical pipe from being climbed. (Photos: Lochrin Fencing Division, The Bain Group Ltd, 15 Aubrey Avenue, London Colney, St Albans, Herts)

an effective disincentive to all but the most determined criminal. To prevent persons getting under the fence, the bottoms of the chain-link panels must be anchored to the ground or set into concrete footings, or the panels may be linked to ground-beams. If the area is not concreted, a concrete apron may be installed either inside or outside the fence line; for example, the soil adjacent to the fence line to an extent of about 1 m by at least 300 mm deep may be dug out, mixed with cement and water to form a

The defended site and its perimeter 25

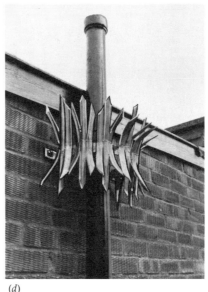

(c)   (d)

weak concrete before being restored. In practice it is found that a mixture of one part of cement to six parts of typical soil is sufficient, and even mixtures of up to 1 in 10 have a significant hardening effect.

It has been found that where a determined attack on the fence can be expected, concreting-in the bottom of the fence can provide added support to the mesh material, so aiding the efforts of the intruder in attacking the bottom edge with a chisel or axe. Embedding the fence to the same depth in gravel is a good method of thwarting this type of attack, in that the chisel or axe tends to bounce off the mesh. One could argue that where this type of attack is anticipated, chain-link fencing would be the wrong choice in the first place.

2-2-8 Metal palisade fencing. This type of fencing is specified in BS 1722 Part 12[3] and consists of pointed flats of sheet metal, configured to give resistance to bending, which are riveted or welded to metal arris rails supported on metal or concrete posts. Such fencing gives considerably greater strength than chain-link fencing. Palisade fencing is more difficult to cut, scale or tunnel under than other forms of fencing, and is very durable. The cost of the installation might be more readily justified if the fencing is to form part of a permanent security installation and the risks of attack are fairly high (see Figure 2.4).

Because of the weight of the fence, the supporting posts must be set into massive concrete foundations to ensure that the fence will not overturn in high winds or on impact by a heavy vehicle.

High-quality galvanised metal palisading has good resistance to corrosion, but is usually painted. Preferably the outer face will be painted a light colour (to make it more difficult to see into the site from outside at night), but the inner face will be painted black or dark green (so that its luminance at night will not impede outward vision).

26   The defended site and its perimeter

2-2-9 For sites having high risks of intrusion, additional barbed wire or barbed aggressive tape may be added to any type of fence. In addition to conventional fencing, various kinds of man-barriers may be erected. These include ditches (water filled, or filled with coils of barbed-wire). It is important that members of the public outside the perimeter are not put at risk of injury; thus, defensive ditches etc. must always be sited on the inside of the perimeter fence which restrains the public from walking onto the defended site unless an additional low fence is placed on the public side of the ditch.

If there are insufficient mobile resources available to constantly patrol a long perimeter, an aid to guarding would be an intruder-detection system (see 2-5). A system that is especially valuable for such applications is a type of barbed aggressive tape which contains an optical-fibre alarm filament. The optical fibre is constantly traversed by coded pulsed light signals; thus, if the tape is cut, the interruption of the light signal chain – even very briefly – will trigger the alarm. A technical intruder could not cut and re-join the fibre quickly enough to overcome the alarm system. Further, deliberately weak sections are provided in the tape, to ensure that if an attempt is made to climb over the fence, the weak sections of the tape will part, and the breaking of the optical fibre will trigger the alarm.

In designing the security plan for premises, one of the important steps will be to determine where the fence-lines shall run – both the perimeter fence and any internal fencing used to create zones (see 2-3-1), citadels (see 2-3-2) and redoubts (see 2-3-3).

It is simple enough to rub out a line on a plan and move the proposed fence a few metres to improve the lines-of-sight or to eliminate a 'dead spot' which cannot be supervised; it is very costly to move the actual fence once it is installed. In order that the topography can be properly taken into account, it would be a sound idea to visit the site and mark out the difficult sections of the proposed fence line with stakes and twine, and then to study the sight-lines of the attackers and defenders.

2-3 Fence-line layout, zoning, citadels and redoubts

2-3-1 Fence-line layout

The design of fence lines must take account of the rise and fall of the land, for this can affect the lines-of-sight for supervision. For example, if there is a deep dip in the land at the perimeter, part of the fence may be out of sight from the security hut or patrol path; this weakness might be overcome by running another fence parallel to the outer fence, placing the second fence on higher ground.

Similarly, it sometimes happens that the Local Authority may impose conditions relating to Planning Consent which include the planting of trees or shrubs within the defended area. These may prevent proper supervision of the perimeter fence or of an inter-zonal fence. In such cases, an additional fence may have to be erected on the inner side of the obstruction (see Figure 2.5).

If there is a building close to the perimeter fence-line, there will be a 'dead zone' behind it, and the stretch of fence that is concealed will be

The defended site and its perimeter   27

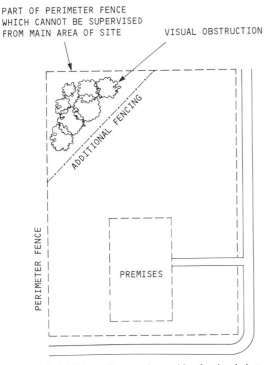

Figure 2.5 Additional fence on inner side of a visual obstruction

vulnerable to being surreptitiously breached. Such a dead zone must be fenced-off (see Figure 2.6), and it may be necessary to provide local security lighting or fit fence alarms (see 2-5). If a building has windows opening into a dead zone (even if properly fenced off), it will be advisable to brick them up. Dead zones must be given special protection, and the patrol path should be arranged to take account of them (see Figure 2.7); further, standing orders for the security staff should instruct they specifically look into and check these vulnerable areas.

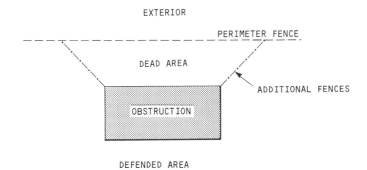

Figure 2.6 Enclosing a dead area close to the perimeter fence

28 The defended site and its perimeter

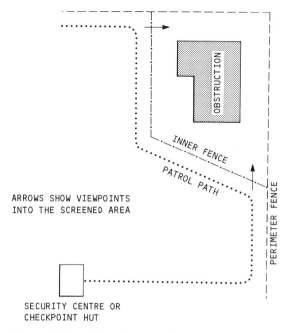

Figure 2.7 Supervision of a screened area

2-3-2 Zoning

The concept of zoning refers to the idea of dividing the total site into several cells by internal fences, and arises from the fact that not every legal entrant to the site needs to have access to every part of the premises. If persons and vehicles can be constrained to remain within authorised zones, the problems of supervision become simpler and the opportunities for crime are limited. For example, visiting transport need not have access to the whole site; office workers need not have access to the works area etc. A tactical control results, and occupants in one zone cannot easily gain access to sections of perimeter fence facing into other zones. The zones must be kept intact, i.e. the fences kept in a good state of repair, and goods should not be stacked close to the zoning fences. In a few critical situations, it may be necessary to provide double fence-lines to demarcate the zone boundaries.

As an example of the need for zoning, the case may be cited of a large factory complex in the West Midlands which has eight entrances for vehicles. Before zoning was introduced, it was possible for a vehicle to enter at a gate legally, and to leave by any other gate. Unfortunately, this freedom meant that visiting vehicles could be driven to a remote part of the site where illegal ballast (in form of bricks, old steel joists, or containers of water) which had been concealed in the vehicle would be dumped, and an equal weight of stolen goods taken on board with the connivance of dishonest employees on the site. The resultant losses amounted to several tonnes of high value goods per month.

There may be gates in the fences which divide the site into zones, but movement between the zones will normally occur only with the knowledge of the security guards. If these inter-zone gates require constant manning the cost may be high. If the zoning design is well thought out, the need for inter-zonal movement may be small. The movements might be served by vehicles leaving the site and re-entering by the gate appropriate to the zone it is required to visit. This results in each such movement being double checked by the security staff.

The application of the zoning principle applied to internal movements within buildings is discussed in section 4-11.

2-3-3 Citadels

The concept of creating 'citadels' for a second line of defence within a defended zone is as old as the ancient arts of building castles and defensive earthworks in earlier centuries. The idea is commonly employed to increase security at military establishments and at national economic targets such as nuclear generating stations, and is applicable to commercial and industrial sites. Just as it is logical to keep valuables in a safe, so it is logical to keep the safe in a secure building; and, similarly, it is also logical to enclose the secure building within a secure fence, and – if the risk is high – within a second line of fencing. A citadel is simply a fenced area within a larger fenced area and not extending out to the perimeter of the site, and is created to give special protection to a vulnerable point ('VP') (e.g. to protect the nuclear zone on a power station site). The inner fence line must have all the necessary features of physical construction, lighting and supervision needed to prevent the entry of unauthorised persons from the outer defended zone into the citadel (see Figure 2.8). Conceivably, special risks might require that a site has more than one set of concentric fences within its boundary, so that an intruder would have to breach three or more fences to get to the VP. Also, a large site might have two or more VPs, each of which could be fitted with an individual citadel defensive system.

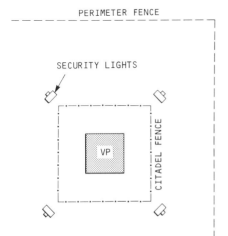

Figure 2.8 Protecting a vulnerable point with a citadel fence and lighting

30 The defended site and its perimeter

2-3-4 Redoubts

Citadels can sometimes serve the function of a 'redoubt'. In ancient times, a safe place was created within a castle, often cunningly concealed, where the king or baron could lock himself in and remain concealed if the castle was taken by an enemy. The leader could hide there until rescued, and might well survive if the occupation was not of too long duration. The concept has some modern applications, for in situations where there is a high probability of the perimeter fence or the structural premises being subjected to determined attack (e.g. sensitive defence establishments, and premises liable to political or extremist attack), in addition to fitting some form of perimeter alarm system and security lighting, it may be necessary to provide physical obstacles to the approach to the inner defence line such as double fence-lines, or ditches filled with coils of barbed aggressive tape. Similarly, safe inner places of retreat can be provided in buildings for the protection of staff subject to attack in cash-handling areas (see 4-4-4, 4-4-5). A suitable area about a vulnerable point may be so protected, or a highly protected area may be constructed solely to provide a refuge for the defenders in the event of an uncontrolled invasion by aggressive persons. A redoubt should be fitted with means of signalling for assistance.

Applying the principle of the redoubt to his own protection, the owner of a professional company who often has to work alone on his premises has fitted his external door with means of identifying callers before admission; but, should he find himself threatened, he can retire into an inner room which is fitted with a strong door and slam-bolts. From within this redoubt he could telephone for assistance, and also has a 'panic button' to operate the intruder alarms. The strong inner door is fitted with an optical door viewer so he can establish if it is safe to come out.

2-4 Exterior security lighting

2-4-1 When specifying the measures for security protection for any kind of premises, the need to provide exterior security lighting should be carefully assessed. We can define exterior security lighting as outdoor lighting provided about premises and operated from dusk to dawn every night of the year with the prime objective of increasing the security from intrusion. The subject is well documented. A convenient introduction to the subject is given in the booklet *Essentials of security lighting*[4]. A treatment of the subject in greater depth is given in the author's book, *Exterior Lighting for Industry & Security*[5]. A more technical treatment of the subject appears in the author's chapter *Security Lighting* in the book *Developments in Lighting – 2*[6].

2-4-2 A brief resume of the subject of external security lighting is as follows:

- *Capability* Security lighting systems provide enhanced night security by (a) deterring the intending criminal; (b) revealing him before, during and after the attack; and (c), in many cases, concealing the defenders from the view of the attacker.

- *System* Comprises (a) physical defences (fences, walls, gates etc.) to slow down the attack and make it more difficult; (b) defenders (security guards, occupants, police etc.) to observe and to respond when an attack occurs or is threatened; (c) security lighting. Any form of lighting will have some security value, but that designed for the purpose will be more effective, more reliable, and cheaper to install and run. Sometimes lighting having an important security function will appear to be provided only for the purposes of amenity, safety, publicity or decor.
- *Experience* Since 1970, thousands of installations in the UK, USA and many other countries. Effectiveness endorsed by military and civil security experts, by police forces and by the Home Office (UK); approved and sometimes specified by insurers.
- *Cost-effectiveness* Comes from (a) ability to supervise large areas and long perimeters with small defending force; (b) reduced need for stringent physical defences; (c) long life of the equipment, high reliability and low maintenance and running costs compared with other more sophisticated systems.
- *Philosophy* It is better to have a security system that positively discourages the crime, and which actually aids the handling of an intrusion situation, than to have a sophisticated system to ring a bell to inform you that an attack is taking place. Lighting is a positive asset when dealing with an intrusion, and is clearly a deterrent; thus it is always better to have the lights on all night than to turn them on (manually or automatically) when an intrusion-detection alarm is triggered. Triggered lighting (trip-lighting) can easily be abused and outwitted by intelligent villains.
- *Physiological factors* Security lighting techniques use proven science-based concepts involving understanding of the process of vision at low lighting levels, the behaviour of the dark-adapted eye, and the phenomenon of glare. Moonlight levels of general lighting are always better than 'Colditz' (beam searchlight) installations or the use of handlamps when patrolling, both of which are heavily weighted in favour of the attacker.
- *Techniques* Five basic techniques are used, which are based on traditional lighting methods and capable of assessment as to effectiveness. They are:

  (a) **perimeter lighting** (lighting of the fence line or boundary, commonly with outward facing luminaires arranged to cause glare to the intruder);
  (b) **checkpoint lighting** (lighting to enable persons and vehicles to be checked at a point of entry or egress; this may include special lighting within the checkpoint hut and special window arrangements to enable the security guards within the hut to see out without themselves being visible);
  (c) **area lighting** (lighting of ground spaces from luminaires mounted on local structures or on poles or towers);
  (d) **floodlighting** – of buildings, plant, walls etc. (to reveal persons locally within the lighted areas, or to reveal them by silhouette against the bright surfaces viewed from a greater distance);
  (e) **topping-up** (the provision of lighting in areas which otherwise would be shadowed by canopies, deep doorways, narrow spaces between

buildings etc.) Modern techniques of topping-up include the use of self-contained engine-powered mobile lighting units (trailerlights, jenny-lights) to deal with temporary risks, hostage situations etc., as well as for paramilitary situations, e.g. riot control, frontier defence. Mobile lighting units are also extensively used on construction sites for providing both light for working and for security purposes (see 5-5).

- *Terminology*:
  - illumination = act or process of lighting something;
  - illuminance = quantity of illumination (measured in lux);
  - lux = unit of illuminance, equal to one lumen per square metre;
  - lumen = unit of light output from a lamp etc.;
  - glare = subjective sensation of discomfort accompanied by a degree of disablement of vision due to the eyes being unable to adjust to excessive brightness ratio in the field of view or in serial exposures;
  - luminance = measurable attribute of brightness, measured in candelas per unit area;
  - luminosity = non-measurable subjective attribute of brightness, the magnitude of which depends upon the subject's adaptation level at that instant;
  - luminaire = a lighting fitting of any kind, containing a lamp or lamps;
  - power = rate of consumption of energy, measured in watts;
  - efficacy = rate of conversion of electrical power into visible light, measured in lumens per watt.

- *Survey* To study the risks, target areas, physical environment etc as a preamble to designing a security lighting installation. Investigations should include method of manning, hours of working, climate, topography etc. Examine perimeter fences from outside and from inside, by day and by night. Keep all notes, plans etc highly confidential, as they would be valuable to a criminal. Protect yourself when surveying by wearing appropriate clothing, safety gear, etc, and obeying all safety and security rules on the site.

- *Applicability of security lighting* Applicable to all kinds of premises, irrespective of size or content; installation details will differ greatly according to location and circumstances, and nature of threat. Darkness is generally the ally of the criminal, not the defenders.

- *Closed-circuit television (cctv)* (see also section 4-8) Security lighting and cctv combine excellently; but sophisticated low-light-level cctv, infra-red imaging, ultra-violet imaging etc., all have disadvantages of high cost and low reliability, plus fact that when an intrusion is detected the security lighting will have to be switched on. Better to have the security lighting on all the time as a deterrent and means of supervision, and use cctv of normal sensitivity.

- *Types of lamps* Generally, the most useful lamp for security lighting installation is the Type SON high-pressure sodium lamp – it has the highest efficacy of all the possible types of lamps (125 lm/W or more), reasonable colour-rendering, long life (5000/7000 hours or more). Do not confuse these with low-pressure sodium lamps which give slightly higher efficacy but have no colour-rendering whatsoever. Tungsten-halogen lamps (type

TH) are used for smaller installations for lower initial cost, but have lower efficacy (only around 20 lm/W), shorter life (2000 h) and thus much higher cost-in-use. Fluorescent lamps (type MCFU) are not suitable for floodlighting etc., but are used in small bulkhead luminaires. Ordinary filament lamps (type GLS) are inefficient (9 lm/W) and have short life (1000 h), and are fragile. High-pressure mercury lamps (type MBF) and metal-halide lamps (type MBI) are also sometimes used for security lighting.

- *Getting advice* In the UK, occupiers can get advice from the Crime Prevention Officer (CPO) of the local police (security lighting is included in the training of these officers, but only at a tactical rather than technical level), and occupiers can consult the industrial or commercial sales engineer of the area electricity board. Some reputable manufacturers provide a free scheme service, as do some larger electrical contractors. For complex, large, or high-risk situations, it may be wisest to seek the services of a security consultant or specialist lighting consultant. The Chartered Institution of Building Services Engineers (London) maintains a List of Consultants. Note that it may be difficult to compare alternative proposals from lighting and installation firms unless an 'outline security lighting specification' has first been compiled by a competent person.

2-5 Alarm systems

2-5-1 In this section we review briefly the technical details of intruder-detection and alarm systems for buildings and perimeter fence systems. Fence alarms are not usually employed for the protection of ordinary industrial and commercial premises as other, less sophisticated and less costly methods are more suitable, and their installation and operating costs are more in line with the risks. For a perimeter fence alarm system to operate successfully, not only must it be highly reliable and free from a propensity for giving false alarms, but the occupier must provide a highly efficient and well-trained security force to react swiftly and in strength when an alarm situation is signalled. It is for these reasons that the use of perimeter fence alarm systems is limited mainly to military and national strategic targets.

2-5-2 As discussed in section 2-2, if there are sufficient mobile resources available to respond in a short enough response time in the event of intrusion being detected, the bulk of the supervision of a long perimeter can be done by the use of some form of intruder-detection system. However, it is unsound to rely exclusively on electronic supervision of fence-lines, for criminals will soon learn that there are no patrols or only very few, and – by technical trickery or co-ordinated attacks on several points around the perimeter – will confuse or outwit the guards.

Nor is it good tactics to use 'trip lighting' (see 2-4), for this allows an intelligent crook to research his crime accurately. By deliberately triggering the lights he can soon learn all there is to know about the defences; such as how many guards there are, if they come on foot or vehicle, from which direction they will come, and how long they take to arrive. If several criminals collaborate, they can have the guards dashing

34   The defended site and its perimeter

from point to point around the perimeter and thus 'wear out the reflexes' until the guards simply cease to respond; they may reach such a condition of exhaustion and confusion that they will turn off the alarm system on the excuse that it is faulty. Criminals can also use instant-response and short-delay audible alarms in the same way.

2-5-3 Various physical phenomena may be monitored to detect intruders. The following are brief summaries of the common detection methods in use:

- *Continuous wiring* Breaking of a wire triggers alarm. Conductors can run along fences, and may be fitted with fragile joints which break upon pressure; wires can be run in brickwork, plaster or concrete; conductors can be run across window openings (typically in fragile tubes which cause open-circuit or short-circuit if tube is bent or cut) (see 3-6); or foil conductors may be fitted to glass; door-opening can be detected with suitable switches (usually reed-switches operated by moving magnets); continuous-wiring systems can be overcome by bridging two points in a circuit and then cutting between the bridging points.
- *Vibration* Alarm systems can be equipped with vibration-detectors. These may be small sealed chambers containing mechanical elements which make or break electrical contact if subjected to movement. Such detector units may be fixed inconspicuously on chain-link fences or any other kind of fence, to give an alarm if any attempt is made to break through or scale the fence. By tuning the detection device to specific frequencies or ranges of frequencies, slow movements (such as may be caused by vibration of the fence by wind) can be given minimum response power and so reduce the risk of false alarms. Other detectors (known as seismic detectors) can be buried in the soil or even embedded in concrete paving, and will detect earth-borne vibrations due to a person moving over the surface or tunnelling under it. Vibration-detectors can also be set into solid walls. Sophisticated versions are claimed to analyse vibrations and not to trip for benign causes.
- *Capacitative detectors* Fence alarm systems may use capacitative cable along the fence, placing the cable in positions where it is almost impossible to scale the fence without distorting the cable or coming into proximity with it. The capacitative cable is tuned to a detector circuit; any distortion of the cable, or the proximity of a new object (i.e. the intruder), changes its capacitance, and the detector circuit triggers the alarm circuit. Prone to false alarms if set to high sensitivity – e.g. would be tripped by piece of paper blown by the wind.
- *Weight* This principle is applied in specially-constructed mats which contain two systems of conductive elements separated from each other by a short distance. Treading on the mat brings the conductive elements into electrical contact and triggers the alarm circuit. Such pressure-mats may be placed inconspicuously under carpeting and other floor coverings.
- *Acoustic* Detectors using normal audio frequencies have limited value in security applications as they are prone to false alarms – e.g. from the noise of a vehicle aircraft, though they are of value in places which should normally be silent (e.g. in a bank vault when locked); can be tuned to be

responsive to selected frequencies or patterns of frequencies, e.g. sounds of breaking glass, splintering wood etc.
• *Ultrasonic* Similar to acoustic, but uses sound frequencies well above the range of human hearing, these being generated by small devices which produce hypersonic vibrations, the reflection of which can be used to detect the presence of a person. Unsuitable for use with fences, but can be used at doorways, in passageways and cable tunnels, etc. They operate on a Doppler effect, and can be triggered by turbulent air (air-conditioning starting up), or random high frequencies (scraping of chair on hard floor in an adjacent room).
• *Light* Systems of detection using visible light beams are not employed along fence lines, but they can be of value within buildings to detect the opening of a door, the passage of a person, etc., but if the light beam does not consist of pulse-coded light, they can be overcome by shining a light into the detector unit. There is increasing interest in the use of coherent light (from a laser), which is capable of being collimated into very narrow beams which can be projected over very great distances, and trip detector devices which trigger the alarm circuit if the beam is interrupted; these could conceivably be employed for defence of an outdoor perimeter.
• *Fibre optics* In another method of using light, barbed steel tape incorporating a glass-fibre optical element is laced through or along the top of a chain-link fence. If the tape is cut, or the tape parts at one of its designed weak points, the pulses of light do not reach the monitoring device and the alarm circuit is triggered. The material can be laced through holes in palisade fences to prevent removal of a pale.
• *Active infra-red* Usually of the beam-breaking type, where interruption of beam triggers alarm; beam can be redirected by mirrors to zigzag across room or doors, etc, but usually limited to only three reflections because of attenuation. If simple infra-red, system can be freaked by hand-held heat-source; more sophisticated version uses pulse-coded infra-red radiation, but this has the disadvantage of attenuating sharply with distance.
• *Passive infra-red* Detects body heat of intruders. Cannot be used for perimeters. Nuisance tripping can occur from heat in motorcar headlamps shining into premises, central-heating system starting up etc. Sensitivity varies with clothes worn by intruder and ambient temperature (see Figure 2.9).
• *Automonitored cctv (see 4-8)* Electronic method of monitoring cctv in which any significant change in the static picture signals an alarm condition. Camera can be framed down to fit a small part of the angular field, e.g. a doorway. With computer control system will scan pictures from a series of cameras rapidly, and will detect any change since picture last seen. Will be tripped by small live creatures, moths etc, flying across picture.
• *Doppler/radar systems* A rather costly but highly effective alarm method is the use of centimetric radar and microwave detector systems. These project a sharply-defined radiofrequency beam, the obstruction of which triggers the alarm circuit. Systems can be applied to fence lines, and also to monitor external open spaces and internal volumes. Claimed capable of being set not to trip from benign causes.

(a)

Figure 2.9 (a) The 'Apollo' passive infra-red detector unit; (b) exterior lighting switched on automatically by the Racalite Light Control System on the approach of a visitor or an intruder. (Photos: Racal-Guardall (Scotland) Ltd, Lochen Industrial Estate, Newbridge, Edinburgh EH28 8PL)

(b)

2-6 Providing facilities for patrolling and surveillance

2-6-1 A factor often overlooked when setting up a security system for premises is the provision of facilities for patrolling and surveillance. An elementary example: at one new high-tech factory, security guards had to don gumboots before doing the routine patrols around the grounds, simply because no paths had been provided for them. The principle to be applied should be that the security staff should be able to move safely (and, if possible, inconspicuously) to every part of the defended area, and should be able to position themselves for clear lines of sight to all vulnerable parts of the premises and any fencing surrounding it. In some cases, vantage points may be provided, in the form of watchtowers, access to roof areas, or simply the provision of windows at the right places for supervision of external areas.

2-6-2 At many sites, because the gatehouse is the only security centre it has also to serve as the security command post. For larger sites employing a number of security staff, and particularly if there are several entrances, this is not usually the best arrangement. It is usually far better to locate the security command post as near as possible to the centre of the site, to give the shortest average distance from the command post to all points on the perimeter, thereby facilitating rapid response to perimeter alarms. Placing the command post well within the site also reduces the opportunities for the intending criminal outside the site to know very much about the security staffing level and the disposition of the staff.

2-6-3 If the security command post (which will contain the alarm panels and cctv monitors, and serve as the communication centre for radio and telephone contact with guards and other staff) is located centrally in the site, the equipment in the gatehouses can be simplified. The main function of the security guards at an entrance will then be to deal with the control of entry and exit at that entrance.

2-6-4 The communication routes within a building should be studied to arrive at the best possible arrangement of the night patrol routes. For example, it might be decided to provide doorways between areas which are normally locked off from each other, such doorways being provided solely to enable the night security guard to patrol the greatest possible area with the minimum route length.

2-6-5 To facilitate supervision, it may be possible to provide small windows or observation hatches in internal walls, so that the security guard can look through into the adjacent areas during his patrols. Such hatches may be fitted with covers, or fitted with optical viewers, so the guard can see through without being seen from the other side. This principle of the security guard being able to see without being seen should be followed through in all its applications, and especially in the application of security lighting (see section 2-4).

2-6-6 It is sometimes thought that the provision of cctv removes the necessity for patrolling. This is untrue, for even the most sophisticated screen picture does not give all the information that can be seen with plain vision. The guard on patrol uses his ears, too; and also may notice a smell which indicates fire or overheated plant etc. By touch he can confirm that doors and window catches are securely fastened. Often the greatest value in having cctv is to enable the guard's colleagues in the security centre to monitor his progress on the screen, as he patrols, to ensure his safety.

2-7 Relationships with adjacent properties

2-7-1 The security risks to which premises are exposed are affected by the nature of the adjacent properties. For example, the number of persons and vehicles attending at the adjacent properties will affect the opportunities for crime at the mutual boundary. The presence of large numbers of persons or vehicles on adjacent roadways or beyond the mutual boundary may provide cover for crimes of opportunity. If part of the boundary to the defended premises adjoins land which is normally unoccupied and not supervised (e.g. farmland, railway land, empty premises), then this may require extra vigilance or strengthening of the perimeter fencing to obtain a satisfactory level of protection. If the defended premises are physically joined to its neighbours, then entry may be effected by breaching the party wall or by entry over or through the roof.

2-7-2 If a common security policy can be operated with adjacent occupiers, mutual surveillance may give greatly added protection. An early-warning system could be devised; for example, permission might be granted to place cctv cameras in the neighbour's area or mounted on his buildings; or a complete mutual defence system (fences, alarms, lighting, patrols etc.) might be developed on a cost-sharing basis by a group of neighbours. At industrial parks, science parks and other groupings of businesses, it is becoming common for the occupiers to share the cost of the main perimeter fence and the lighting of the estate roads. The principle could be extended to include the joint employment of security staff and suitable security lighting and alarm installations for the whole site. Such cooperative systems would doubtless be cheaper to operate than individual systems, and probably would be more effective too.

2-7-3 Having instituted common security measures, the occupiers would do well to participate in the management of site security in all its aspects; for example, a secure and rapid means of communication between those persons in each company responsible for security of their premises will be of great value in case of an intruder crossing a common boundary, or where a crime is committed cross-boundary, e.g. the staff of one occupier committing a crime against an adjacent occupier.

2-8 Control of vehicles; parking

2-8-1 The windows of the checkpoint hut will ideally command a view of all approaches to the premises, and enable the security guard to supervise both vehicular and pedestrian movements (if both occur at that entrance). For large premises, or where considerable numbers of staff enter and leave together, or where vehicle movements are frequent, it will be safer if vehicles and pedestrians are directed through separate channels, and better supervision will be achieved too. Separation permits a single guard to give his full attention either to the vehicle channel or the pedestrian channel, without the risk that persons may slip through while he is dealing with a vehicle.

It is desirable that there shall be inner and outer gates or traffic bars on a vehicle channel, and that the distance between these shall be long enough to accommodate the longest vehicle (plus its trailer) that is likely to enter. Then the gates can be closed ahead and behind the vehicle while it is being checked or searched (see 4-3)(see Figure 2.10).

Large sites should be zoned (see 2-3) to constrain the movement of vehicles to only those parts of the site necessary for their proper business. A one-way system, with vehicles entering at one point and being checked-out at another, may have advantages for traffic control as well as for security. Otherwise, entry and exit lanes may be partitioned from each other by a fence so that the staff can control both lanes and give a high level of supervision. Various layouts are possible, including placing the checkpoint hut between the two traffic lanes.

2-8-2 If vehicles are permitted to enter a fenced site and park within the perimeter, it must be assumed that there is a risk that vehicles will be used as an aid to crime, or that there will be thefts or other crimes committed against the vehicles or their contents. Some organisations try to divest themselves of responsibility by allowing vehicles to be parked on their land but outside their protected perimeter, allowing free and uncontrolled access into the parking area from the public road. If this is done, for legal reasons it may be necessary to display a sign with wording such as 'No public right of way. This land is not dedicated to public use, but visitors to the premises of Bloggs & Co may use this car park at their own risk'. It will not be possible to prevent unauthorised parking, and despite display of the notice, it is doubtful if the occupier could avoid civil responsibility for injury or loss suffered by a visitor (whether authorised or not) if the occupier was negligent, i.e. failed to provide suitable lighting, or allowed there to be a dangerous hole in the road, etc. Allowing uncontrolled public access to the carpark probably frees the occupier from responsibility for losses suffered by visitors due to criminal activity in the carpark or associated with the vehicles using it.

2-8-3 If the vehicle park is inside the protected perimeter of the premises, it will be necessary to control entry and to supervise the area. Vehicle parking areas are sometimes located out of the sight of security staff, and then it may be necessary to install cctv to supervise them.

40   The defended site and its perimeter

To save labour costs, some companies arrange for the control-barrier at the entrance to the parking area to be operated from the reception area, and this requires that the reception area has a window overlooking this, or – better still – a hatch so that visitors can speak to the receptionist. The system is weak, for the receptionist cannot leave her post to deal with matters at the barrier.

(a)

(b)

Figure 2.10 (a) 'Roadblocker' vehicle barrier; (b) torque barrier (Photos: Henderson Security Gates, Kelvin Lane, Crawley, West Sussex, RH10 2ND)

A card admission system (using punch-coded or magnetic-strip cards to raise the barrier) may be an advantage if the parking area is used by the company's own vehicles or staff vehicles. If an electrically operated barrier is used without direct supervision by the occupier's staff, although such equipment is generally very safe, there is a risk that misoperation of the barrier could cause injury to a third party (a child, for example) or damage to a vehicle, for which the occupier might be held responsible at law.

2-8-4 If vehicles – especially loaded lorries – are parked on the site at night, providing a citadel fence around the parking area will reduce risk of interference (including that from own staff). Security lighting will further reduce risks at night, and increase safety. It has been noted that heavy lorries parked at a dockside parking area tended to park closer together when the illuminance was ample – some 13% more commercial vehicles could be safely parked on the site after lighting was installed. (see also 6-6-2).

2-9 Illegal entry facilitated by topography, weather and seasons

2-9-1 Every site is unique, and must be surveyed with care and imagination. In devising the security plan (see 1-8, 7-8), one must first evaluate every feature of the site from the viewpoint of defence against attack. High ground outside the defended area may give a tactical advantage to an **intending intruder**, and this must be negated by reducing his ability to see into the defended zone, e.g. by erecting opaque fencing or walls, or by the use of security lighting that deliberately subjects the outsider to disabling glare at night (see 2-4).

There may be obstructions which prevent observation of the perimeter fence from within the site; obviously, it will not usually be practicable to remove buildings and costly structures, but temporary low-value structures might be removed, or earth-moving operations carried out to remove hummocks and mounds to improve the lines of sight. If an obstruction cannot be removed, it may be necessary to erect an additional fence-line inside the obstruction and to provide vantage points or patrol-path routes that will enable the supervision of the hidden area behind the obstruction (see 2-3).

Open flat land outside the perimeter is difficult to cross undetected by day, and is probably far more difficult to cross undetected at night if suitable security lighting is in use. But if cover is provided by bushes and shrubs – or even by heather or long grass – surreptitious approach may not be difficult under any conditions of light. Man-made cover, in the form of abandoned vehicles, rubbish and tipped spoil, will also give concealment. Note that a ditch will give concealment to an intruder, just as a hummock or bush will do. The cleaning up of any land on the approaches to the perimeter fence, and especially between outer and inner fence lines, should include levelling the land as much as possible, filling-in dips, trenches, ditches or old excavations.

Trees near the perimeter fence may be used by intruders, for an overhanging branch – perhaps with the aid of a rope or ladder, or with access gained from the roof of a vehicle – could permit even a sensitive detection system of the fence to be bypassed. Because there were no trees at the time of the security survey, or because at that time the trees were very small, the matter may be overlooked but, some years later, the trees may afford means of illegal entry. The removal or pruning of trees can significantly improve fence security, as will the use of bollards or high kerbs outside the fence to prevent vehicles parking close to it.

It would be wise to investigate if the land inside or outside the site has any abandoned mineral workings, dene-holes or wells. Ancient tunnels may provide a means of entry, as may culverted watercourses and old sewers and floodwater drains. The locations of such features may sometimes be discovered by study of large-scale maps of the area, or by reference to Ordnance Survey maps, parish records, archives in the local government offices and old conveyances of the land. At high-risk locations it may be prudent to do such researches into the features of adjacent land as well as of the defended site itself. Long forgotten underground workings and tunnels can sometimes be seen by aerial photographic survey. Features that are normally invisible may show up when the area is photographed with infra-red film, or when the filming is done early morning or just before sunset when the sun's rays are more nearly parallel to the ground.

2-9-2 Security guards may be reluctant to make the required number of outdoor patrols under bad weather conditions, and efforts must be made to provide them with suitable clothing and warm rest quarters so they can perform this duty without hardship. Security lighting may be less effective under conditions of fog or mist, and directional lighting (provided to produce a concealment effect) may have to be switched off under bad visibility conditions because of the glare-back from the fog. Heavy rain or hail or strong winds may produce such a noise level that suspicious sounds are not readily heard, and extra patrols at high vigilance may be needed to maintain the required standard of security.

2-9-3 A site becomes more vulnerable if winter flooding permits approach by boat (e.g. inflatable dinghy) to parts of the site that otherwise would be difficult to reach. A bog, lake or river may be crossed easily when frozen. A site becomes more vulnerable in summer when foliage obscures the sight-lines of defenders. In summer, streams may dry up and boggy areas dry out, so that intruders may approach on foot or drive over areas normally impassable.

Chapter 3

The building

3-1 Functions of the structural enclosure

3-1-1 It is not usual for buildings to be designed with security against illegal entry as their prime function. The design features are provided mainly to satisfy other objectives, so that the security engineer later must try to overcome the structural and tactical weaknesses which have been built in. In addition to its security functions, the structural enclosure of a typical building has the following purposes:

- *Weather protection* The structure and its cladding give protection against thermal loss and gain, and against ingress of moisture, the degree of climatic modification depending on the function of the building. For example, if for human occupation, it will normally be protected against damp, radiant heat loss and draughts, and will be capable of being heated for occupier comfort. In contrast, a warehouse designed for storage of bulk goods could be efficient for its purpose and yet provide a much lower standard of climatic modification. In general, modern low-cost walling and cladding materials intended primarily as thermal and moisture barriers do not provide high resistance to criminal penetration.
- *Acoustic insulation* In principle, the best insulation against the entry of noise through a structural plane is the provision of mass; but high mass does not always provide great physical strength. For example a hollow wall filled with sand would have excellent acoustic attenuation, but would be easy to penetrate.
- *Fenestration* Most buildings intended for occupation are fitted with windows which admit daylight, may be a means of ventilation, and provide the amenity of outward view. A high proportion of illegal entries are effected through windows. Adequate protection of windows against entry can be costly, and may be difficult to achieve in a manner acceptable to occupants. Buildings required to have high security may have limited fenestration or no conventional windows at all. A windowless building may be acceptable for some functions if the lighting and ventilation are effective.
- *Privacy* Architectural means of providing acoustic and visual privacy may create a false impression that privacy equates with protection from penetration. For example, in buildings fitted with false ceilings and internally divided into cellular rooms by partitioning, the fact that entry

44   The building

can be gained into any cell via the ceiling void may be overlooked (see 3-7-2).
- *Fire protection* A building having good features of fire resistance may not be physically secure. Further, the provision of means of emergency escape may conflict with the needs of security (see 3-2).

3-1-2 Traditionally, industrial and commercial buildings have been constructed to have a projected life of some 50 years, though many buildings now in use are far older than this. Many older buildings have massive construction and can readily be adapted to provide high standards of physical resistance to criminal attack; but many modern buildings are relatively lightweight structures relying upon a steel and concrete beam framework for strength, and having easily penetrated curtain walling. Wooden-frame or system-built buildings with a projected life of 30 years or less, may return a lower cost per annum than conventionally constructed buildings, but are more difficult to make secure. The extra cost of a more massive construction might be traded-off against reduced costs for the installation and operation of security services throughout the life of the building.

The converse of the foregoing can also be be true; for example, it is possible to build a lightweight structure, and rely upon a system of fences, alarms and supervision etc. to achieve the level of protection required. This was the course followed in the construction of a large bonded whisky warehouse in Scotland, where it was estimated that a saving of nearly £1M was made by erecting a simple frame-structure warehouse clad in weather-resistant building board which would take a criminal only a minute to penetrate. The windowless structure was designed to give only the minimum climatic modification necessary without the normal amenities for occupation, with reliance on the external security system to prevent the approach of criminals (and unauthorised vehicles) to the building curtilage. A factor in calculating the economics in this case was that it was planned to use the site only for 10 years before removing to other premises; but experience showed that the total cost of operation (building maintenance plus cost of operating and manning the security system) was so low that it was decided to continue the occupancy for at least a further decade.

3-2 Access and exit – conflict between requirements

3-2-1 Without special adaptations, a building from which it is easy to escape in emergency will tend to be easy to break into. Emergency exits are often points of criminal entry. A window through which occupants could escape in emergency may offer a route in for the criminal; an external fire-escape staircase or ladderway can provide a convenient means of criminal access and is also difficult to supervise at all hours. Criminals may enter a building legally or by trick; they may then commit their crime and escape via an emergency exit. Systems using centrally-operated remote electric locking of doors (with the electric locks released automatically on operation of the fire alarm or on mains failure) are obviously easily overcome; such systems must be authorised by the Fire Prevention Officer

The building   45

and the insurers. Any system in which all occupants must carry an escape key is not acceptable, and would be easily capable of misuse. The conflict between the need for easy access with the need for restraint upon entry and misuse of exits has to be resolved.

3-2-2 Exit doors can be fitted with alarms, but this is only of value if there will always be a security force able to react quickly. Where an emergency exit gives on to public roadways, the criminal may easily escape before the security staff can attend. An intruder who enters during the day and conceals himself on the premises may leave with his loot via a fire exit. A possible solution to this problem is to 'sweep search' the building at close of operations for the day, section by section, to ensure that no-one remains behind, locking internal doors and then locking all exit doors including emergency exits. Such a procedure may be acceptable to the Fire Prevention authority and the insurers if it can be demonstrated that there is no possibility of any person being locked-in in an emergency; this would involve having a procedure to ensure that all exits are operable before the building is re-occupied. Unfortunately this arrangement is not generally acceptable in the UK following a fire in a large retail store when a number of members of the public died because exit doors had not been unlocked before the commencement of business for the day. However, it is common practice in cinemas and theatres for the panic-bar locks on emergency exits to be chained and padlocked after the public have left the building each day.

3-2-3 A possible way of dealing with this problem would be to arrange that escape routes pass checkpoints which are manned during working hours; or it could be arranged that escape routes from the building lead only into a fenced compound – safe for the escapees, but restraining them from departing completely from the premises. Persons leaving by an emergency exit would have to be released from the external pound by the security staff who would be alerted by the operation of the door alarm.

3-3 The structural shell and cladding

3-3-1 The lightweight construction of many modern buildings has led to many instances of penetration through the walls, the outstanding example being the driving of a vehicle through the cladding sheeting of warehouses in order to break in (see 6-2-3).

3-3-2 Most roofs other than concrete-slab roofs offer little protection against penetration by persons familiar with their construction. Tiles and slates are readily removed, and entry can be effected without creating much noise. Roofing felt is easily cut, and supporting wooden boards prised up. Asbestos-sheet roofs are easily entered by quietly cracking the areas around the fixing-bolts and raising a panel. The noise of breaking through a roof may not be heard at ground level. Some protection can be provided by sandwiching expanded-metal grille material between the roof layers, or inside an existing roof. Continuous-wire alarm systems, proximity-sensitive cable systems, and acoustic-detection systems can be installed in lofts and roof cavities.

3-3-4 All glazing in roofs is vulnerable, especially hollow formed plastic roof-lights which are easily and quietly prised out of their frames. Wired-glass gives no better protection than plain glass; indeed, it may be easier to crack and lift out in one piece. Once access to the roof void has been gained, the criminal can enter all the rooms beneath it (see 3-7).

3-3-5 Buildings may have plant located on the roof, such as air-conditioning and heat-pump equipment, humidifier or de-humidifier plant, air-handling equipment, and even stand-by generating plant. In association with such equipment, there may be apertures in the roof for trunking and ducts, and external ladderways provided for access to the plant for maintenance. Any duct of greater diameter than, say, 250 mm should contain stout bars to prevent entry. Consideration should be given to control of access to the roof (see 3-3-6), control of movement from the roof area to the protected parts of the premises (see 3-3-7), and supervision of roof areas (see 3-3-8).

3-3-6 Premises may be attacked via a route including the roofs of adjacent buildings, and access may be obtained by breaking-out upward through a roof or through a party-wall at roof-void level. Security of existing buildings may be improved by barriers (e.g. railings, barbed aggressive tape) to impede access to the roof. Following the precedent of US practice, some buildings in the UK are now being fitted with external escape staircases having a sliding counterweighted bottom ladder section which automatically descends when the escapee puts his weight on it. These are easily used for access by the pulling down the counterweighted section, or by an insider bringing it down to admit an accomplice. An alarm switch which is operated by movement of the counterweighted section could be fitted.

3-3-7 Staircases and ladderways into roof-voids should be locked off to prevent access. Any roof-hatch or door should be of robust construction, and should be kept locked (if this is not incompatible with safety or fire requirements) and preferably fitted with an alarm device. External pipes could be treated with non-drying anti-climbing paint.

3-3-8 Supervision of all approaches to the roof, internally and externally, may present considerable difficulties in the case of large premises and those consisting of groups or complexes of buildings. Note that access may be possible from one roof to another by climbing along pipe-bridges and cable-bridges between buildings, or by laying a plank or ladder across the gap. Illumination of the roofs will enable supervision by direct vision, security mirrors, or cctv scanning. If there is a high block or tower structure higher than other buildings in the complex, this could be used for visual observation or cctv to supervise the roofs.

3-4 The floors and foundations

3-4-1 When attempting to apply security disciplines to an existing building, it is easy to be deceived by the apparent solidity of the structure and believe that penetration would be so difficult that intrusion would not be attempted by cutting through masses of concrete and brick. Unfortunately, there have been many cases of intruders breaking into buildings by the simple expedient of gaining access to a cellar or basement, and then breaking upward through the wooden planking of the ground floor. This points to the need to fortify doors and windows which give access into a basement, even if it is not occupied or used for storage, and to provide appropriate defence by alarms.

3-4-2 Even solid floors may not have the resistance to penetration that the degree of risk dictates. For example, many industrial buildings which are not designed for heavy floor loads may have a surface layer of small-aggregate concrete which is no more than 50–60 mm thick, and this is laid on a bed of ballast or shingle topped by 20–30 mm of sand. This means that an intruder who can tunnel under the perimeter foundations would find little resistance to breaking upward through the floor to gain access to the building.

3-4-3 Many buildings are constructed on the foundations of a much older building, and it may well be that sewers and land-drainage culverts have not been effectively blocked off, so that penetration may not be difficult to an intruder with experience of the building industry. In recent years there have been several spectacular major robberies at banks where the thieves have broken into the vaults by tunnelling from nearby main sewers. There is usually no difficulty at all in gaining access to sewers; one simply lifts a man-hole cover.

3-5 Doors

3-5-1 In discussion of windows, we noted that breaking through a window is the most common route for a forced-entry; for non-forced entry, the door is naturally the most common route for the criminal into the protected premises. In a high proportion of cases, the criminal simply walks in (see 1-3). Doors of strong construction may readily be opened if one has illegal possession of a key (see 4-10), or if someone responsible for security has simply omitted to lock up (see 7-10).

3-5-2 It is a foolish economy to fit a good door with a cheap lock which may easily be picked or forced. Doors which are themselves of adequate robustness may fail to give the resistance required because they are fitted into flimsy doorframes, or are hung on hinges of inadequate strength. The strength of a good lock is wasted unless it bolts into a really secure frame. Commonly, the doorframe is not sufficiently well anchored to the structure and can easily be prised out of position with a stout jemmy, frame and all.

3-5-3 Doors serving the function of providing an emergency exit may need to be fitted with panic-bolts permitting instant opening from within only. Note the conflict between the requirements for escape and for security (see 3-2).

3-5-4 A careful analysis of the movement requirements may reveal that a building has more doors in use than is necessary for convenience and safety. The security manager could decide to seal permanently any such redundant doors (preferably by bricking-up) to reduce the inherent security risks, but this is an action upon which the Fire Prevention Officer should be consulted first. Doors not required for emergency use, and other doors which can be locked when the premises are not occupied, could be protected with an external steel roller shutter (see 3-6-3).

3-5-5 Some dwellings and business premises have recently been set on fire by malicious persons who have poured petrol through the letterbox and ignited it. It is unfortunate that the idea has spread among criminals, political and racial extremists, persons involved in industrial disputes and the violent young, so that having a letterbox in any external door must now be regarded as imposing a serious risk. Alternative actions and precautions will include to seal off the letterbox and either:

- Take out a Post Office Box Number and arrange for the mail to be delivered only during working hours or collected daily from the postal sorting office by a representative of the organisation, or
- Fit a new letterbox to a gate or wall that does not give directly into the premises. The new letterbox is to be backed with a weatherproof and locked letter container inside the gate or behind the wall.

3-6 Windows

3-6-1 The disadvantages of windows

All windows present some security risk, even if non-openable, if they have an aperture large enough to admit an intruder. If called upon to assess the worth of windows on a scale of security values, most designers would prefer to erect a windowless building, or one in which windows had been reduced to vestigial vision-slits provided only for the enjoyment of the occupants and through which no intruder could pass.

If writing the specification for a new building, or when putting up proposals for reducing the size of or eliminating existing windows (proposals which may be vigorously opposed by others), the security manager or consultant may need to take a broader view of the matter and consider the disadvantages of windows in factors other than security. The following summary of the negative aspects of building fenestration may be of value:

- If the building is to be air-conditioned, a windowless structure will be cheaper to erect and maintain. The capital cost per square metre of window area may be ten times higher than for plain curtain walling, and

The building   49

the annual maintenance cost throughout the building life will be higher too.
- Windows are sources of glare and adventitious light (light that does not enhance visibility for the occupants). In deep-plan offices having perimeter windows, it may be necessary to use a higher illuminance of electric lighting during daylight hours in order to offset the glare effect of the windows, with greater energy usage and cost.
- Windows are a source of adventitious heat by insolation (sun heat radiation). The solar heat-gain through windows will add to the air-conditioning load of the building.
- All through the life of the building, the heat losses through windows must be made good by heat input to the building. It can often be shown that relying mainly or totally on electric lighting for daytime illumination will provide a higher standard of visual comfort for the occupants and can be justified economically; indeed, the waste heat generated by the lighting installation itself may be sufficient to provide most – if not all – of the heating requirement of the building, leading to significant economy in energy usage and cost.
- Draughts from opened windows or from ill-fitting windows can disturb planned flows of force-draught or air-conditioning ventilation.
- In interiors where high standards of cleanliness are required, openable and unsealed windows permit the entry of dust, and chemical and organic pollution as well as insects.
- Finally, windows admit noise which is difficult to suppress (especially that due to nearby trains, heavy or fast road traffic and passing aircraft).

3-6-2 The benefits of windows

The benefits of providing windows in buildings are mainly related to human enjoyment and the creation of a pleasant environment. It is possible to design windows which are constructed, positioned and shielded so that they give minimal nuisance from the admission of unrequired light, heat, and noise, which have small heat losses and no draughts, and which give the building's occupants the enjoyment of natural lighting and a view of the outside world. Systems of lighting to augment the natural lighting may be used, e.g. 'permanent supplementary artificial lighting'. In some buildings, openable windows serve the function of access for firemen, or as smoke-extract apertures in fire, or may be used as a means of escape in emergency.

3-6-3 Security measures applied to windows (see also 1-8)

Although every window in the structural enclosure of the defended building is a potential hazard to security, it is possible to design windows so as to make them difficult to enter. Grilles and protective bars may be applied, though this course is not favoured as tending to give a prison-like appearance.

Barring of windows is often ineffective because the bars are too flimsy. Tubular bars (e.g. gas-piping or electrical conduit) are sometimes employed, but are of very little use. However, it may noted that in one system of intruder alarms, light-section aluminium-alloy tubular bars are used inside the glazing to carry an internal tensed alarm wire; bending or cutting the tubular bar will trip the alarm, and the appearance from outside is as of a strongly barred window (see 2-5-3).

Barring a window opening may be ineffective because the bars are not robustly seated into the surrounding structure. For effective protection, bars to windows must be of steel or cast-iron, at least 20 mm diameter, and must be braced by cross-plates or other means every 300 mm or so. The maximum spacing between bars should be 120 mm. Both ends of the bars should be grouted with hard cement into the surrounding structure, a penetration of 100 mm being the minimum required entry into concrete. When sinking window bars into brickwork, the bars should be set at least 55 mm back from the face of the brickwork, and preferably should be a 'brick and a half' back, i.e. set back by 160 mm. Penetration of the bars into brickwork should be at least a 'course and a half', i.e. not less than 130 mm. If these guidelines are followed, a bar cannot be prised out by merely splitting one brick.

An ancient security device seems to have been rediscovered recently, namely the use of **external window shutters.** One modern form of this excellent idea is a steel roller shutter which is wound down from inside the building and cannot be opened from outside. This provides a very high degree of physical security. Extra precautions can include fitting alarms to the shutter mechanism, and providing means to lock the mechanism. For large windows (and doorways), electrically driven shutters may be employed.

3-6-4 Window frames

The preferred criminal method of entry through a window is to force the window frame rather than incur the risk of being heard smashing the glass. Most lightweight window frames can be quickly and quietly forced with a suitable jemmy, against which the addition of domestic-type window locks gives little protection. Some extra resistance to forcing is provided by having a deeply-inset hardwood frame made in one piece (and not having a separate nailed-on external trim-strip which is easily prised off). PVC window frames have virtually no resistance to forcing; aluminium window frames of lightweight section do not offer as much resistance as well-made steel frames. Double-glazed windows, especially of large area (including patio doors), are not as rigid as they look, and can be warped out of their runners or frame by attacking one corner with a jemmy. A common fault with window security is that the window frame may be secured to the structural brickwork by just a few thin nails, and the casement or sash which carries the glazing may be removable from the frame with minimal force. A usual method of retaining the glass is by a few brads on the outside, these being concealed by the fillet of putty or glazing compound with which the fitting is made weathertight. Putty etc. can be cut out from outside and the brads extracted, so that a pane of glass can be quietly removed by a skilled intruder. Some modern window designs have extruded plastic strips to seal the glass into the frame, and it may be possible to prise the strip out of the recess from outside and quietly remove the pane of glass without breaking it.

The building   51

3-6-5 Window locks and security

As well as by the use of direct force, entry into premises through windows may be effected by sabotage of the defences in various ways, e.g. by interference with locks or hinges of windows during the day to prepare for later penetration after the building has been locked up. For example, window staybars (of the kind which serve the dual purposes of propping the window casement open and securing it when closed) can be interfered with, e.g. the staybar can be nearly sawn-through, or the screws which secure the staybar or its latching-plate can be loosened. This sabotage can be done from inside the building, or from outside when the window is open, and enables easy and silent entry at some future time. Countermeasures to this kind of entry mode will include:

- use of steel-framed windows in preference to wooden ones;
- peening-over the heads of securing screws to prevent them being loosened;
- applying robust key-locks to all openable windows (even if they are protected by internal or external security bars).

3-6-6 Glazing

There is a popular misconception about the security value of wired-glass. It should be understood that the common types of wired-glass (for example, 'Cathedral Wired Glass') give no greater protection against entry than does float glass of the same weight. Indeed, if the pane is forced by any means, the internal wire mesh actually gives the intruder the advantage that he will not have to stick brown paper with glue or treacle over the glass (to reduce the noise and danger to himself) before smashing it. However, there are some special glasses which incorporate an alarm wire cast into the plate. In another product, an alarm wire lattice is sandwiched between two layers of glass which are cemented together, but again, there is little if any gain in mechanical strength.

The following notes give some information about common glazing materials, many of which are covered by British Standards:

- *Float glass* This is the common glazing material for windows. It may be noted that glass becomes more brittle as it ages. Lightweight window glass that has been in situ for 20 years or so will fracture on pressure much more readily than modern float glass.
- *Plate glass* This is glass of good optical transmission quality which is generally used for shop windows and for glazing of larger windows. When mounted in a heavy glazing frame which fully supports all edges, it is difficult to smash except by the application of considerable force.
- *Toughened glass* This is plate glass which has been toughened by a heating and annealing process. Toughened and laminated glass (consisting of two sheets of glass sandwiching a thin membrane of tough clear plastic material) has considerable resistance to impact, and the glass tends to hold together when cracked.
- *Bandit-resistant glass* This is heavy gauge laminated glass of strength that will resist considerable impact.

52   The building

- *Bullet-resistant glass* This is a heavy gauge laminated glass of high strength and fitted usually with a further layer of tough clear plastic on its inner surface to prevent spalling of glass fragments when struck by a bullet or missile on the outside. Such glass is used for security screens at banks etc. Important note: there is no such thing as 'bulletproof glass'; any bullet-resistant glass would be penetrated by a bullet or missile of sufficient mass and velocity, though grades are available which will absorb the energy of a bullet or shot from specified types and calibres of weapons at specified distances.
- *Plastic sheet glazing materials* For use instead of glass, clear or tinted polycarbonate sheet of great mechanical strength is available, which having flexibility, is not easily broken. Retained in a deep frame from which it cannot be sprung, it presents a formidable obstacle to impact or forcing. However, it is much more costly than glass. It may be damaged by the application of heat, and in service may be marred by scratching.

Figure 3.1 Windows of Local Authority dwellings in the City of Glasgow glazed with 'Meshlite', a glazing material consisting of expanded-steel mesh in a sandwich of polyester resin and glassfibre. (Photo: Mitra Plastics, Mitra House, Whittington Road, Oswestry, Shropshire, SY11 4ND)

The building   53

- *Plastic cellular glazing panels* These are used for glazing rooflights. They have the advantages of light weight and good thermal properties as well as high light transmission; but they represent a serious threat to security as they are easily cut with a sharp tool or prised from the frame. They generally do not have resistance to high temperatures, but tend to distort and fall out of the frame when subjected to rising hot air from fire (which can be an advantage if it is desired for the rooflight to act as a smoke vent).
- *Adhesive films for application to glass* Plastic films having various qualities of light transmission, colour, radio-screening etc are available to be applied to glazing in situ. Certain grades are claimed to give resistance to intrusion or create greater safety to occupants if the glass be broken by a missile etc. Applied to ordinary glass they do not provide greater resistance to penetration unless the edges of the glass are deeply set into a robust frame, and preferably the adhesive film will be applied to both sides of the glass.
- *Glass-fibre reinforced plastic sheeting* This has great mechanical strength and resistance to impact, but clear grades are not available. The sheeting may be reinforced by being moulded around a layer of expanded-steel mesh which is thus completely embedded into it; used in small panes within a strong window construction (of the kind in which the pane cannot be removed from outside), this material allows a fairly high proportion of incident light to enter, but provides good resistance to damage or penetration (see Figure 3.1).
- *Glass bricks* These are usually hollow, of clear glass but obscured. They are used to admit light to stairwells, toilet areas and passageways etc. They are laid with a special cement into structural wall apertures. If of suitable grade and properly fitted, they offer about the same resistance to penetration as common bricks.

3-7 Interior walls and partitioning; ceiling voids

3-7-1 The modern demand for flexibility in floor plan layout has led to the construction of many open-plan buildings in which the floor areas are divided off by lightweight partitioning. This is acceptable when one organisation occupies the whole floor, but is unsatisfactory in the case of multiple occupancy buildings where the partition wall between two businesses could be dismantled by an intruder in a few minutes by use of an ordinary screwdriver.

3-7-2 Many modern buildings have a cavity above a false-ceiling, the cavity housing various building services and distribution systems. Typically a ceiling void may contain heating and ventilation trunking, hot and cold water supplies and connections for the fire sprinklers; lighting equipment; conduits and cables for lighting, power, fire-alarm systems, telex and computers – and possibly the wiring of the alarm system. Because of these many obstructions, it is not usually possible to carry the partition walls up through the suspended ceiling to the structural ceiling, though a minimum number of fire-break partitions will be installed. Some 2-hour duration

54   The building

fire-break partitions are effective against flame and smoke, but can be breached with a screwdriver in a few minutes. Once a thief gets into the ceiling void, he can open further ceiling panels and drop down into every open-plan or cellular office on that floor. Indeed, if the fire-break partitions in the vertical service ducts are also flimsy, an informed thief could make his way about the entire structure without even having to open a door.

3-7-3 A further disadvantage to security of having a ceiling void is that it provides an excellent place of concealment for an intruder (and one that would hardly ever be searched). The ceiling void also is somewhere to hide stolen goods until the thief judges it is a safe time to remove them. In one case, an intruder gained access to a multiple-occupancy office block by presenting a business card which was not his own, and, once out of sight of the receptionist, he made his way to a toilet area where he was able to get into the ceiling void through the roof of a toilet cubicle. Later, when the building emptied, he broke into several firms on that floor by dropping down through the modular ceiling, and then lowered his loot from a window to his waiting accomplice. He spent the night in the void, and the next day made his way out of the building unnoticed. This could be a blueprint for many crimes of this kind if suitable precautions are not taken by building occupiers (see 6-1).

3-8 Bridges, walkways, tunnels and external facilities

3-8-1 In the practical management of premises, it is easy to overlook the various ways in which buildings in a complex are connected together to provide means of mutual access. The objective of this section is merely to alert the security manager or consultant to the possibility that such features of premises may present a problem area that must be carefully studied.

3-8-2 In the case of older premises, bridges, walkways at roof level, tunnels and other connections between buildings may have been installed piecemeal over a long period to provide answers to problems of access and communication as the organisation grew; conversely, bridges and other structural connections between building elements are important architectural features of many modern industrial and commercial complexes. It is not unusual, for example, for buildings which are separate at ground floor level to have a common basement area or underground carpark; also, buildings which are **conjoined** but operationally separate may share fire-escape arrangements in the form of external staircases or roof walkways. Always, in studying the security risk posed by such arrangements, one must resolve the conflict between the needs for access and exit and the needs of security (see 3-2). In many modern complexes, several occupiers may share facilities for access and escape, car-parking etc., and hence have a common interest in the security aspects (see 2-7). In investigating these matters, take note of the possible presence of disused underground tunnels or culverts (see 2-9-1).

3-9 Electrical services, power supplies and standby power

3-9-1 Electrical services

Because the correct operation of the electrical services in any business premises is vital to the functioning of the business and the maintenance of required standards of security and safety, the security manager should maintain close liaison with the building manager, departmental managers and engineers responsible for the maintenance of the premises and its electrical services. The following points may be relevant:

- All plans to carry out engineering works affecting electrical services should be discussed with the security manager well in advance to ensure that there are no unexpected interruptions of power supplies to security installations, lighting etc. (see 5-5). Where such interruptions must occur, the security manager can prepare for them, e.g. by providing additional personnel, or hiring-in standby power plant or temporary mobile lighting etc.
- Interruptions to communications services can be particularly disruptive; prior warning will enable temporary telephones to be installed, or radio links set up etc., before the normal installation is taken out of service.
- Trunkings and cable-ducts carrying power cables may also carry telephone circuits, computer cables and alarm circuits. If contractors are to have access to these, it may be necessary to vet the personnel who will be involved, and to provide special supervision.

3-9-2 Power supplies

In the UK it is usual for industrial and commercial organisations to take their electrical power from the public supply system; however, self-generation of part or the whole of the load may be economic, particularly if the demand is excessively peaky, or if there are steam-using processes, or if full use can be made of the waste heat from diesel engines driving the generators.

The significance of this in terms of the security plan for the premises lies in the possible vulnerability of the generating plant to illegal interference, requiring that it be housed and supervised at a level of security appropriate to the risk. For practical reasons (noise, smell, vibration etc.) generating plant may be in a fairly isolated position on the site; this may require study to reduce risks, e.g. consider additional fencing of this vulnerable point (see 2-3), local security lighting (see 2-4), local alarm system, cctv (see 2-5), and particular attention by security personnel (see 4-7).

3-9-3 Standby power

It is standard practice for alarm systems and supervisory systems (e.g. cctv) to be provided with standby power in the form of reserve batteries to enable them to operate during a power outage. Small items of equipment may have replaceable primary-cell batteries, but the more common arrangement is for the batteries to be of a rechargeable type which are kept in a good state of charge automatically. The period of reserve provided by batteries is usually only a few hours, and for a system which will remain fully operative during longer periods without the normal power supply some form of secondary supply will be required. There are special requirements regarding the available duration of battery power for emergency lighting (see 3-10-3).

One way of achieving greater security of supply is to have two intakes from the public supply system, these being derived from different feeders (the electricity board can advise if this is possible for any particular premises). More commonly the possibility of failure of the public supply is covered by the installation of reserve batteries or generators.

Reserve batteries of massive capacity have the disadvantages of high costs of installation and maintenance; they require good ventilation to limit the risk of combustion of gases given off by secondary cells while charging. Battery systems for emergency lighting and other services may have a central battery, a zonal battery, or the equipment may be fitted with local or integral batteries (see 3-10-3).

Batteries store energy in a form that issues only as direct current (d.c.), but the battery output may be converted to alternating current (a.c.) by rotating machinery (motor-generator set), or by static inverter. The former will be more commonly used for heavy loads, the latter for light loads.

Standby generators and motor-generators are commonly arranged to start up automatically on failure of the mains supply. Because instantaneous start-up of rotating equipment is not possible, there may be batteries with static-inverters etc. to provide power to bridge the interval between mains failure and the start-up of the rotating plant.

3-10 Interior lighting and security lighting; pilot lighting and emergency lighting

3-10-1 Interior lighting for security and safety

The interior lighting of the premises is an important factor of security and safety, for the security staff will need sufficient and suitable light inside the building to patrol in safety and to maintain surveillance of the premises without the use of torches or handlamps. This is necessary so they can give the proper standard of protection against intruders and unauthorised acts by or movements of staff, and for vigilance against fire. Electric lighting will be needed even during daylight hours for patrolling windowless areas, basement floors, enclosed stairwells etc. where daylight does not reach. There is a Common Law duty imposed on all occupiers to ensure the reasonable safety of persons on their premises, and this includes for the provision of light. The Health & Safety at Work Etc Act[7] requires that 'sufficient and suitable lighting' be provided for safety and to enable the emergency routes to be used.

3-10-2 Pilot lighting

For larger premises, three systems of lighting may be employed:

- *Normal lighting* This is provided for the safe movement and performance of normal working tasks by the occupants. The electric lighting may be in augmentation of any available natural lighting.

- *Pilot lighting* The amount of light needed for patrolling and supervision will be far less than that needed for the normal occupancy activities, so a system of pilot lighting may be employed for economy and convenience. This could consist of selected luminaires in the normal lighting system being switched on to provide the necessary subdued and economical level of lighting for patrolling and supervision, but is more likely – for practical reasons – to be a system of small luminaires specially installed for the purpose. A useful safety feature is to have the pilot lighting come on automatically when the normal lighting is switched off during normal working hours. This provides for safety during lunchtimes, tea-breaks etc., and at the start and end of the working shifts. One method of providing pilot lighting is to use 'maintained escape lighting' in the emergency lighting system (see 3-10-3).
- *Emergency lighting* This is a system of lighting provided to enable persons to escape in any emergency when the normal lighting (or pilot lighting) is not working because of failure due to any cause (mains failure or local system failure or damage, etc.)(see 3-10-3).

The use of the three systems of lighting is shown in Table 3.1.

Table 3.1 Use of lighting systems

	Normal	*Pilot*	*Emergency*
Normal working hours	On	Off	Off
Non-working hours; building patrolled	Off	On	Off
Building closed; not patrolled	Off	Off	Off
Mains failure any time*	Off	Off	On

* It can be arranged that if there should be a failure of the mains when the building is not occupied, the emergency lighting circuits will not be activated.

3-10-3 Emergency lighting

Emergency lighting should not be confused with stand-by lighting which is provided to enable work to continue during any failure of the normal lighting, e.g. by a mains failure. Emergency lighting as defined by BS 5266 Part 1: 1975 & amendments[8] is lighting for escape from premises. Its function is to save life by enabling persons to escape from any actual or threatened danger – even if the normal lighting has failed. A common danger is that of fire; thus every attempt should be made by features of the design of the equipment and its installation to ensure that the emergency lighting will continue to function for as long as possible in a **conflagration**. Personnel and premises are exposed to special security risks during any outage of normal lighting, and thus the security manager should regard the installation and maintenance of a suitable system of emergency lighting as meriting a high priority in his duties.

CIBSE Technical Memorandum TM 12 – Emergency Lighting[9] gives guidelines on the method of designing escape lighting systems, while ICEL specifications 1001[10], ICEL 1002[11] and ICEL 1003[12] issued by the Industry Committee on Emergency Lighting give technical guidance on the construction of emergency lighting equipment and the design of installations, and form the basis of a Certification scheme for this type of equipment.

The objective in these installations is to provide sufficient illumination to enable persons to escape from the premises – an illuminance level of approximately that of full moonlight (0.2 lux) being typical.

It is usually a condition of the Fire Certificate issued in respect of UK premises that an emergency lighting system is provided, and that the battery capacity shall be such as to give a required duration – e.g. 1, 2 or three hours – to enable all persons to escape, and for proper searching of the premises for casualties to be carried out. If the batteries in an emergency lighting installation should be inadvertently partially discharged, so that the duration remaining was less than the specified time, it would be illegal for the premises to be occupied until the batteries had been sufficiently recharged as to give the required minimum duration.

As part of the proper methods of maintenance, the batteries of any emergency lighting system (central batteries, zonal batteries, local or integral batteries) must be tested for capacity by being discharged and recharged. Recharge time is usually at least 8 hours, and so this operation has to be carried out on occasions when the premises are not in use – e.g. weekend, night-time, summer shut-down, Christmas holidays, etc. However, other methods have to be used to charge/recharge the emergency lighting batteries at premises that are constantly in use, e.g. airports, power stations. In some cases the batteries are removed and replaced with fully recharged and tested ones; in other cases a portable battery is installed temporarily; alternatively a mobile generator set has to be brought to the premises and operated during the discharge/recharge cycle on the installed batteries.

Chapter 4
Supervision and control of entry

4-1 Setting the objectives for the system

4-1-1 If the entry-control system is to make an effective contribution to the security of the premises, a set of rules must be laid down for the performance of the duties of the security guards. These rules must be rigidly applied, without latitude to adapt or relax the rules in particular cases; for, if the guard has the power to relax regulations or exercise discretion, undoubtedly he will do this, and the resultant standard of security will be lower. If he can allow irregularities (even if these are reported afterwards) then it must be assumed that in the fullness of time he will either (a) be tricked by a dishonest person, or (b) conspire with others to permit or facilitate a crime.

4-1-2 The objective of carrying out searches on persons and vehicles entering and leaving the premises is usually primarily to discover evidence of theft or other crime. In particular cases, search of incoming persons and vehicles will be necessary to search for weapons, explosives etc., being carried in connection with a terrorist attack or violent crime. Search may be made for illegal objects, e.g. for cameras and tape-recorders being carried into a secret/secure area, or matches or smoking equipment being carried into a fire/explosion-hazard area, as well as for tools which could facilitate a breaking crime, e.g. safe-cutting equipment.

4-2 Practical difficulties of operation

4-2-1 Security personnel guarding large complex premises with several entrances should guard against variations of the 'two badge trick'. This consists of the first villain presenting a convincing story at reception, and being issued with a pass badge. The villain then surreptitiously leaves the premises without surrendering the pass badge, and re-enters at another reception point where he is given a second pass badge. He then contrives to convey the second pass badge to a conspirator (e.g. dropping it to him from a window), so that there are soon two villains free-roaming the premises.

As an example of how easy this trick is, the author managed to collect no fewer than five pass badges at a large petroleum installation by entering by each entrance in turn, and then leaving by walking through an emergency exit door which was neither locked nor alarmed (see 3-2).

Protection against infiltration of this kind can only be achieved by carefully checking the credentials of all visitors, and ensuring that they are accompanied at all times until they are off the premises and the head-count adjusted. Exit doors may be fitted with alarm switches so that persons leaving by unauthorised routes are detected by the security staff (but see 3-2).

4-2-2 An example of practical difficulty encountered with entry-control systems concerns a northern industrial works visited by the author where a costly admission-control system had been installed. Employees enter the building by presenting their magnetically-coded cards to the reader-slot, which not only opens the gate but also automatically computes their time record and wages. But, during the author's visit, it was noticed there was considerable traffic through a side door leading through the security office. A man who wanted to go back to his car to fetch something was allowed to go out and return ten minutes later – thus showing up a fundamental weakness in the control of entry. The admission-control system was first class; but the control of admission was very slack.

4-2-3 If a security system is to work efficiently, the security staff must have clear authority to deal with situations where crime has been carried out or is suspected; this may involve questioning members of the staff and visitors, searching vehicles and possibly examining personal luggage. Difficulties may arise if the requirement to co-operate with the security staff is not written into the employee's employment contract. He may have sound legal grounds for refusing to open his bag or submit to a body search if no such contract condition exists. In general, it will be found that visitors tend to be quite co-operative as regards opening up briefcases on entering a building, for this has become a routine matter since the world-wide rise in terrorist attacks. Strangely, some visitors who have had their bags examined on entry, may object to submitting to a further search on leaving.

Except at defence establishments, visitors are unlikely to agree to being 'patted down' (checked for weapons), let alone being strip-searched, unless the searching is carried out by a police officer.

4-2-4 Notwithstanding the correctness of their actions, company security staff may be accused of bullying, interference with employees' rights and freedoms, or of 'acting like the Gestapo' when they insist on checking parcels and bags carried by employees. Indeed, in the UK there have been strikes following the discovery of stolen goods in an employee's possession when searched – the feeling of solidarity between employees tending to override their commonsense understanding that the employer simply cannot permit uncontrolled theft.

In large organisations, where the contact between the shop floor and the management takes place mainly or solely through trade-union representation, it is essential that every opportunity is taken (at meetings of works liaison committees, for example) to make it clear to the employees' side that their interests – as well as the company's interests – are served by preventing attrition of the company's assets by theft. Just as there must be liaison on matters such as industrial safety, management must seek the co-operation of staff in all aspects of security.

4-3 Checking vehicles

4-3-1 Checking of vehicles has the objectives of:

- ensuring that the driver is identified, and that his papers are in order;
- ensuring that any person accompanying the driver is authorised to be in the vehicle, and that the driver is not under duress;
- ensuring that the vehicle contains only legitimate cargo, and does not contain 'false ballast' (e.g. paving-stones under the driving seat, or plastic cannisters of water concealed under the chassis) which can be dropped off and replaced with an equal weight of stolen goods during the visit to the protected premises – possibly with the connivance of an employee (see section 4-3-2);
- under certain conditions, and possibly mainly at **politically sensitive premises,** ensuring that the vehicle does not contain weapons, explosives etc., bearing in mind that these might be carried without the knowledge of the driver.

4-3-2 Searching a vehicle at a checkpoint can be a time-consuming operation, leading to queues and delays if not executed efficiently. Under pressure, the security guard is likely simply to wave vehicles through, or decide to inspect only a proportion. He must be able to see into high trucks, and preferably should be able to see on to the roof of any high-topped vehicle. For this, wheeled steps may be used, or a convex security mirror placed at a suitable point.

Search under vehicles will be facilitated by the use of a convex security mirror on a long handle, or perhaps on a low wheeled trolley with a light. A successful way of getting light under a vehicle is to bounce it off a light-coloured road surface, say a coating of hard-wearing white epoxy paint. This can be hosed to keep it reasonably clean, and adjacent lighting can be directed appropriately.

4-3-3 It is probably impossible to entirely prevent weighbridge frauds, particularly if there is collaboration between the driver and a dishonest employee of the occupier. The general pattern of these crimes is as follows: a vehicle from an outside organisation or a carrier company arrives at the gate with a load on board, there usually being a considerable number of packages. Each package may be identified by a name or number, and is listed with its weight on a way-bill carried by the driver. The weight of the load (as listed on the way-bill) plus the tare of the vehicle agrees with the total weight of the vehicle as measured on the weighbridge. When the vehicle departs, its weight will have increased or decreased by the weight of the goods delivered or collected, so the weighbridge check gives no grounds for suspicion. But, the security guard has no way of knowing if the way-bill details are correct; he has neither the authority nor the time to cause the vehicle to be unloaded and all the items to be checked and weighed. He does not know if any of the packages are dummy loads. The only steps he can take to prevent false ballast (see 2-8-4) being exchanged for stolen goods (usually with the connivance of a dishonest employee) are as follows:

- as far as time and facilities allow, inspecting the vehicle carefully, and operating the weighbridge according to rules (including making careful records);
- meticulously checking documents and weights as far as is possible, including checking the weights of those parcels delivered or collected at the premises;
- thorough and continuous supervision of the visiting vehicle, ensuring that the driver or his mate have no opportunity to drop false ballast (e.g. by discharging water or sand etc.), or to pick up unlisted parcels;
- if the site is zoned (see section 2-3), exercising responsible operation of internal gates and barriers to restrict the opportunities for crime.
- the alert security guard will have his suspicions aroused if he finds unexplained weighty materials dumped in the lorry park or on a quiet part of the site, e.g. sand, stones, scrap metal etc, or if he finds unexplained pools of water which might be due to the dumping of water ballast.

4-4 Checking of visitors – prevention of walk-in crime

4-4-1 Unauthorised persons entering the premises

Many buildings that are made very secure when they are unoccupied at night are vulnerable to theft during the day simply because little effort is made to control entry of persons.

At large premises such as hotels, office blocks and hospitals, walk-in criminals may pass themselves off as legitimate visitors or members of staff. Appropriate uniforms or overalls are easily acquired, and it is unlikely that intruders will be challenged if they move about the premises in a purposeful manner. Many such intrusions occur, and doubtless the losses and disruptions would be much less if ways of preventing this easy illegal access could be found.

Badge and pass systems only work well if they are constantly supervised, and therefore carry a high labour cost. Electronic pass cards to operate door locks can be used (see 4-10), but usually it is not practicable for busy staff to repeatedly check in and out when passing through control points within the building.

A spot check on the persons using a staff dining area during the night at a hospital revealed that quite a few unauthorised outsiders were regularly using the premises to gain the benefit of subsidised meals. Anyone can walk around hospital corridors and look like a patient or a visitor.

4-4-2 Thefts of portable equipment

Because of the small size and high value of modern items of office equipment (plus the ease with which they can later be sold to unscrupulous buyers), offices are often the targets of walk-in thieves. Costly small items may be spirited away by spurious 'maintenance mechanics', or may be found missing after a visit by a strange window-cleaner or by an inconspicuous chap who came to measure for new carpets or to estimate for fitting extra electric sockets. Too often, security is slack in dealing with visitors, who – after being attended to by a member of staff – may be left to find their own way out of the building. Having made a seemingly normal call, the visitor is left free to wander about the premises looking for something to steal. An intruder may make a legitimate visit to Company A to create the opportunity to take a purposeful stroll round the offices of Company B who happen to share the building.

Equipment thieves may be highly skilled, and act with audacious calm and charm when challenged. 'Sorry, I think I am in the wrong office!' they may say when found in a room alone. The methods of theft include caching the stolen goods on back staircases (or even in another firm's office!) for later collection.

4-4-3 Thefts from staff

An informal investigation by the author produced surprising results as to the value of cash and valuables that people bring to their place of work. A quiz of staff in a sales office revealed that most of them usually carried £100 to £200 on them – usually in a wallet which was left in their pocket when the jacket was hung on a peg or left on the back of a chair – even when they went out of the room. In a solicitor's office, several secretaries wore watches and jewellery estimated to be equivalent in value to a year of their salaries.

When personal possessions are stolen at a place of work, much ill-feeling and suspicion is generated, though in fact habitual thieving from workmates rarely goes undetected. Pilfering of handbags, wallets etc. is a profitable quick crime for walk-in thieves. Cash that is extracted from a wallet or handbag cannot be identified.

A common pattern is for the thief to take the handbag or wallet to the toilet, and to leave it there after removing the cash or valuables. Quite often, stolen items such as credit cards, cheque-books and savings books are concealed on the premises for later retrieval on a future visit when it will be safer. This means that the pilferer is 'clean' if searched after the time of the crime. A favourite hiding-place for small items is inside the flush-cistern of the lavatory, the goods or cash being protected with a well-knotted polythene bag. Other popular caches are on the tops of cupboards and partitions, or on outside window-sills, perhaps even hanging down from the sill on a piece of string so as not to be noticeable from inside the premises.

4-4-4 Thefts of cash

Visiting a large publishing house, the author was asked to wait in the entrance hall, and was seated in front of a notice which indicated the direction of the wages office with an arrow. Interested, he followed the arrows, and moments later walked into an unlocked office where girls were counting a large quantity of bank-notes just collected from the bank. Comment on this incident is unnecessary.

The methods of protecting wages offices and cash-handling departments are well-known and can be highly effective; but there are many incidents of robbery from such areas where elementary precautions were not carried out. It is a sad reflection on a management if it cannot protect the staff from the dangers of a robbery. The safety of staff is of far higher importance than the mere protection of money.

4-4-5 Countermeasures

The countermeasures for combatting the general risks of walk-in crime should include:

- Create a strict barrier between the protected area and the area which is accessible to the public. Do not permit staff to use fire-exits as the route for slipping out to buy cigarettes. All visitors must enter only through the reception route and, having been identified and their business verified, should then be passed into the supervision of a member of staff.
- No visitor should be left unattended except perhaps in a restricted area such as a waiting-room or the outer part of a reception area. He should be conducted by a member of staff at all times, and escorted off the premises politely at the end of his visit.
- Verify the identity of visitors, particularly those coming to maintain or inspect equipment, or to carry out window-cleaning, building repairs etc., discreetly keeping them under surveillance at all times. Someone working on a task, e.g. fixing carpets or maintaining a photocopier, may wish to visit the toilet, or may need to return to his van to get tools; make sure he does not wander off into unattended rooms, nor carry anything out of the building without full explanation.
- Costly items of equipment such as facsimile machines and electronic typewriters can be secured to the desk by screwing or padlocking, or may be fixed with 'sticky-bar' devices which, while holding the item securely attached to the desk-top, can be released with a key for maintenance purposes.
- Apply an indelible security mark to all valuable items to enable them to be identified if stolen and later recovered by the police. The presence of the mark is a positive discouragement to theft, though secret marks (visible only under ultraviolet 'black' light) are often used. The mark need only consist of the initials of the company and the post-code. Keep a safe record of the descriptions and serial numbers of all items which might be targets for theft.
- Keep unattended offices locked. A pass-key should be held by a responsible person for use in case of urgent need to open the room.
- Partly-glazed doors and glass partitioning will aid supervision of an area. If privacy is needed (e.g. in the personnel department and in directors' offices at certain times) blinds can be fitted.

Supervision and control of entry   65

- Keep an accurate record of all visitors, including their time of arrival and departure. Check that all visitors have left at the end of a working day – searching the building if necessary. Car registration numbers of visitors should be recorded in the visitors' book – that is, the observed number should be written down by a member of staff (rather than asking the visitor to record the number of his car which is out of sight of the receptionist).

The countermeasures to the risk of thefts from members of the staff should include:

- **Training of staff,** with frequent reminders to take proper care of their own valuables.
- **Provision of lockable drawers or lockers** for staff personal use, with strict key control.
- **Seek full staff co-operation** in the supervision of visitors.
- **Immediate and thorough investigation** of all claims of loss of property by staff, including always notifying the police.

The countermeasures to combat the risk of thefts of cash should include:

- Encourage staff to accept payment of wages by credit transfer or cheque so that cash transactions in the building are limited.
- If it is necessary to dispense cash on the premises, create a teller's cubicle with a no-grab till-basket and glazed with bandit-resistant glass. The teller's cubicle should be entirely enclosed within a security screen and have a stout door that can be locked from the inside. Provide a safe in the teller's cubicle so he or she can place cash within it before opening the door leading into the adjacent cashier's office.
- The cash-dispensing window should be located where its approaches can be supervised by other staff; it should be located with care, and be neither in a quiet out-of-the way corridor nor too easily accessible from the street. Preferably the cash-handling area will be located so that it is approached through two or even three sets of doors which are to be opened only one set at a time. Consider use of electrically-controlled door-locks or turnstiles (see Figure 4.1), with means of interrogating visitors before admission to the cash-handling areas (speech only, or with cctv)(see Figure 4.2).
- Consider installing an entry-control system using coded electronic admission cards or tokens, plus a personal code or password (see 4-10).
- Provide physical barriers (e.g. steel doors) to prevent quick entry by 'blaggers' (strong villains with sledge-hammers).
- Install alarm-buttons which can be swiftly and unobtrusively operated by staff when under threat. In some cases it may be preferable for the alarm system to be fitted with a delay before the sounding of the local bell or siren.
- Provide wages staff with means for them to sweep cash rapidly into a safe. Also provide them with a 'redoubt' – this being a strongly-protected inner room where they can take refuge in an attack, even if this means abandoning the cash (see 2-3-4). Protection of staff from threat or injury must be of paramount importance.

Figure 4.1 'Rotostyle Model RS' turnstile for control of entry. This is operated by an access control sytem such as the use of a coded card in combination with keying-in an identifying number. (Photo: Mayor Turnstiles Ltd, Mayor House, Station Road, Edenbridge, Kent, TN8 6HN)

Supervision and control of entry 67

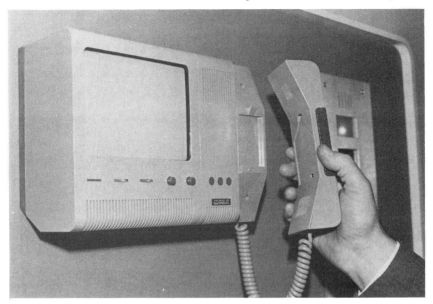

Figure 4.2 A video entryphone to control entry into a building or to regulate movements of persons between areas within a building. (Photo: Henderson Security Electronics, Unit 4, Tannery Road Industrial Estate, Gomm Road, High Wycombe, Bucks, HP13 7EQ)

4-5 Checking of staff; personal searches, vehicle searches

4-5-1 Checking of staff

It is a generally-accepted custom for hourly-paid employees to clock in and out of their place of employment, and it has long been established in the courts that any falsification of the time records (e.g. by not attending while clocked-in, or by one person punching the clock-card of another) is fraudulent. It is normal custom for security personnel to oversee the clocking procedures of other employees. Where the numbers are large, employees may be required to carry a means of personal identification. Modern systems of time-recording use magnetically coded cards which give identification, act as magnetic keys, and generate entry and exit data which is fed directly to the computer for calculating the pay of the employee (see 4-10, 6-5-5).

4-5-2 Personal searches

Persons entering politically sensitive premises and military establishments may be asked by police or security guards to open their baggage for inspection with an explosive-sniffer device, and may also be asked to submit to personal search. Because most members of the public are used to search procedures when passing through airports, we tend now to accept such as being necessary. Personal searching for weapons may be carried out electronically (the subject steps through a detector frame, or is run over with a hand-held metal-detector), or the subject may be 'patted down'

68   Supervision and control of entry

– a procedure most people find rather objectionable. Only police and customs authorities have the right to carry out intimate body searches. A visitor to a business organisation could be asked to open his briefcase, and if he refused he could be excluded from the premises.

In the UK, an employer does not have the right to search an employee entering or leaving the premises – nor to search his car – unless it has been made a condition of employment that he shall submit his person and his vehicle to be searched when required. If an employee refused to be searched, this might give rise to suspicion (even if submitting to search was not a condition of his employment). If searching is to be carried out, careful attention to procedure will be necessary, the searching to be carried out in a secure and private place, with at least one witness to the actual search. Except when performed by police officers acting correctly under laws and police procedures, searching of persons and vehicles can only be carried out with the agreement of the subject.

When a security guard stands watching employees leave the works, he has only a few seconds to scan each one, and he may easily miss the suspicious signs of bulging pockets, large objects concealed under coats, or lunchboxes which seem uncommonly heavy.

Familiarity breeds laxness, and the guards may also feel that stopping and searching staff could cause a breakdown in goodwill between the security staff and other employees. Indeed, it is not unknown for trades unions to make strong protests against random searching, even when searching reveals stolen goods! The situation is illustrated by a fictional tale about a man who went out through the works gate pushing his wheelbarrow full of straw every evening for twenty years. Sometimes the guards searched through the straw, but never found anything. On the day the man retired, the guard asked him, 'Look, it doesn't matter now, so tell me: what were you stealing?', and the man replied, 'Wheelbarrows!'.

The mass exit of hundreds of workers at the end of a shift may give convenient cover for goods to be taken out illegally in pockets or under coats etc. An example of the magnitude of the problem at a large works can be given, where it was apparent to the management that large numbers of small but valuable components were being stolen by the staff. The route from the department where the suspected thefts were occurring to the gate was enclosed by railings. One evening, the security guards locked the gate just before the end of the shift, so that some three hundred men found themselves stopped in the roadway before the gate. As soon as they were assembled there, the guards closed-off the building doors behind them. Then some announcements were made over the public address system along the lines of, 'Will Security Officers Numbers 12, 17, 23 and 35 attend at once at Gate F, please!' – these being intended to give the impression that considerable numbers of security personnel were going to make personal searches. In fact, no searches were carried out, and after a few minutes delay the gate was opened and the men allowed to leave the premises. When they had gone, over sixty items were found on the ground, these being stolen goods to a total value of about £500.00. After this incident, because the workforce feared that a mass search might be carried out, the frequency of thefts declined resulting in a saving of over £2000.00 per week.

4-5-3 Vehicle searches

If the employer permits employees to bring their vehicles into the premises, it is a reasonable condition that random searching of vehicles may be carried out. Of course, the employee must be present when his vehicle is searched. The employer may make other rules to regulate the situation, for example, not permitting employees to return to the vehicle car park during working hours.

An incident occurred at a research establishment where a notice was displayed at the gate saying 'Cyclists must dismount and wheel their bicycles through the gate'. One evening a departing employee was seen having some difficulty wheeling his bicycle; he lost control of it, and it fell over. The guard watched with interest when he saw the man struggling to stand the bicycle up again, and went to help him. He found the bicycle was extraordinarily heavy, this being due to the fact that the employee had drilled holes in it and had filled the cavity of the tubular frame with over 50 kg of stolen mercury.

4-6 Response to an alarm situation

4-6-1 The certainty of there being a response to an intruder-detection alarm is the principal reason for criminals (a) being deterred from attacking premises fitted with alarm systems, and (b) aborting crimes on the sounding of the alarm and escaping from the scene. If the prime objective were to catch criminals, it would be better if there was no local audible alarm, but that the alarm signal should be transmitted silently to security agencies who would come to the premises rapidly, or who would investigate by scanning the premises by cctv from the remote station (see 4-8) or by listening with sensitive microphones.

Occupiers are more interested in defending their property than catching criminals (see 1-1-5), and they tend to opt for a compromise, with a short delay between the transmission of the signal to the police or security agency and the sounding of a local audible alarm. Most intruder-detection systems send no signal to remote agencies, but merely sound a local alarm.

4-6-2 The response to an alarm can be evaluated by the following factors:

- The confidence level that there will a response. It is quite common for criminals to 'sound the drum', i.e. test for reaction by deliberately triggering the alarm system. This may be done on several occasions, sometimes long in advance of the planned time for the real attack.
- The reaction time (time elapsed between triggering of the detection device and the arrival of the responding force); this would ideally always be shorter than the time it would take the criminal to escape from the scene, which is not possible if the local alarm sounds too early.
- The reaction strength (number and quality of men and vehicles etc. which can be deployed). In the guarding of large premises (as with the policing of a district), one should not send all one's forces to respond to an alarm. There may be more than one attack at different places at the same time; some attacks may be feints to draw the reacting forces away from the real attack. It may be better to get a small number of defenders to the

70   Supervision and control of entry

scene very quickly rather than a larger force a few minutes later. The rapid-reaction force (if only one man) can summon help by radio or other signals arranged for the purpose when he has established the need.

If the prime objective is to arrest the intruder, it would be better if the security agency vehicle or police car did not approach with sirens blasting, but made a silent approach. Conversely it may be argued that the bells and two-tone horns will panic the intruders, perhaps causing them to abandon the loot, or to depart with such haste as to be easily identified on the road.

4-6-3 The response to an intruder-detection system may come from within the perimeter of the premises (response of security staff on the premises), or from outside the premises (by virtue of a telephone or radio signal etc., transmitted to a central station (i.e. a security agency monitoring switchboard) or police station. Systems dependent on telephone lines are vulnerable to criminal interference. If the line is only activated on the triggering of the alarm, a dead line may become apparent only when it is needed. The cost of maintaining a continuous open line through the public telephone system would be prohibitive, but electronic monitoring of the telephone line can be arranged at low cost, so that a cut or shorted line would trigger the alarm. The use of optical fibres transmitting continuous coded signals gives constant monitoring of the integrity of the cable and system (see 2-5-3).

In some cases recently, criminals have opened British Telecom manholes in the street, and cut through all the cables with an axe. This results in making all the telephones and alarm systems in the vicinity dead; but if the essential lines are electronically monitored, the occupier or the security agency could be alerted to bring extra manpower to his premises until the system has been restored. Unfortunately, cable-chopping attacks can take days or even weeks to repair, during which time many premises in the vicinity are vulnerable.

4-6-4 Greater reliability may be achieved by the use of short-wave radio links to send the alarm message to the police via a central monitoring station. Such systems are driven by a non-interruptable battery-powered electrical supply. Electronic knowledge on how to jam or spook radio links becomes ever more common, so that sophisticated coded transmissions are necessary. Another method which recommends itself is to transmit continuously by a line-of-sight ultra-short-wave radio signal (or possibly even by a laser beam or high-intensity visible light beam) to a local monitoring mast from which the alarm condition is relayed to those who will provide the physical response.

4-7 Day and night patrolling, internal and external

4-7-1 Unless directed and managed with understanding, security guards who have to patrol fence lines and buildings routinely may find the job boring, and it becomes difficult for them to maintain a high level of interest and vigilance in this dull task which has to be continued for a large part of their working time. In order to stimulate alertness and interest, various kinds of tests and inspections may be instituted, to prove the system and also as a training method and means of supervision.

4-7-2 When security guards are moving about the site during the day or at night, it is advisable for them always to be in full uniform (including the hat) so they are instantly recognisable. A smartly turned-out security guard gains respect from both staff and visitors, and his **meticulous performance** of all duties is visible evidence that a good security system is in operation – and hence is a discouragement to crime by both staff and outsiders.

4-7-3 When new security staff are being inducted, and especially when training a new team to conduct the surveillance and patrolling in a new security system, it has been found helpful to explain the principle of 'fences, lights and men' to the staff concerned. Security guards who have not previously worked on a site equipped with security lighting may not appreciate the capabilities of the system; therefore the author makes it his practice to provide each member of the security staff with a copy of the following instructions:

Instructions to security personnel – working with security lighting

1 Although it may be necessary for you to carry a torch or handlamp, try to make minimum use of it. Shining your torch when you are on patrol tells any hidden intruder exactly where you are and what you are doing. He knows that if the beam of light from your torch does not fall upon him, you probably do not know he is there, and thus he will have the tactical advantage. Further, when you look at the bright patch of light made by your torch, your eyes become adapted to that brightness, so it becomes difficult – if not impossible – for you to spot a dark-clothed intruder in the shadows.

2 When in the security hut at night, use only the minimum essential lighting. This will enable your eyes to adjust to a lower brightness, so that you will be able to see out of the windows more easily, and, when you step outside, you will adapt more quickly to the low lighting level there.

3 When you go outside at night, give your eyes a few moments to become adapted to the darker conditions, and then you will see far more than if you had switched on your torch. Rely on the security lighting, and do not impatiently start using your torch for ordinary movement about the site.

4 Security lighting works best as a component in a three-part system consisting of (a) the physical defences, i.e. fences, locks, bars etc., (b) good supervision by the guards, and (c) the lighting. The lighting alone will not defend the premises well; neither will physical defences if you are not there; and neither can you defend the premises properly without the physical defences. At night, with the best physical defences and the most efficient guards, the premises would still be vulnerable if suitable security lighting is not provided.

5 Security lighting helps you defend the premises, its contents and the personnel in three ways: (a) by revealing the intruder when he approaches your defended area, (b) in many situations by providing you with concealment behind the glare of the lights, and (c) by a combination of these two factors, to deter all but the most determined villain.

6 Darkness is not usually your ally – it tends to help the intruder. In the dark, on a quiet night, you might hear a pin drop at a distance of two paces. But, with good security lighting, you can see the man dropping the pin at 200 paces.

7 At dusk each evening, check the lighting and check your torches. Report any failed lamps in the security lighting system. Report also any signs of damage to the security lighting equipment. Criminals sometimes try to sabotage the lighting to aid a later attack.

8 Use the lighting tactically, and only switch on your torch when you really need to. Think about how someone might try to break in; what route would they take? What would they be after? When are they most likely to come? Approach each target area so as to surprise any villain who may be there. Come from behind the glare of the lights; conceal yourself behind the lights; give the intruder all the disadvantages of your cunning tactics. The lights are there to help you.

9 Patrol like a fox – never going by quite the same route twice. Vary the times of your patrolling; vary the number of patrols over each part of your patch each night. Come to each vulnerable area from a different direction each time. Stand still and listen for a few moments; listen and watch. Imagine how the lights would be affecting you if you were the intruder.

10 Security lighting must be on all night, every night of the year. If it gets switched off during the dark hours, or is switched on only as needed, it is not true security lighting. Do not try to save a little electricity by switching off security lighting; it is a well-designed system that uses energy wisely and for good purpose.

4-7-4 Experiments have shown that a dark-adapted subject focuses typically at a distance of around 2 m if the visual field is so bland (i.e. is so featureless and lacking in interest) as not to stimulate him to extend his focus. This condition is known as 'night myopia' and was reported by Rayleigh[13]. The phenomenon is of importance in studies of visibility through chain-link fencing at night, for, if the security guard should patrol inside the fence at a distance of about 2 m, he will tend to focus his eyes on the fence itself rather than look through it to keep watch on the surveyed field. The condition is believed to be responsible for the 'sleeping sentry' situation, where the sentry apparently fails to see events outside a fence or railing; some have been found guilty of dereliction of duty (and, in war-time, in some cases, shot!). To prevent the 'sleeping sentry' condition, the following may be done:

Supervision and control of entry 73

1 The brightness of the fence as seen by the patrolling guard should be reduced by:

(a) employing chain-link fencing which is coated in black or dark green pvc, or painting the inner face of palisade fencing with dark paint, so that the fence reflects less light towards the guard;
(b) arranging for any outward facing lights to throw their light over rather than through the fence mesh;

2 The patrol path should be set at a greater distance than 2 m from the inner face of the fence. It is found that the 'sleeping sentry' effect is minimised if the patrol path is greater than 4 m from the fence face.

3 Security guards should be trained in this matter, and be aware of the fact that they may unconsciously look at the fence rather than through it, and should be encouraged to 'stretch their vision' by deliberately observing distant objects through the fence. It may be a helpful idea to plant small white-painted pegs in the surveyed field at a distance of 20 m to 50 m from the fence to act as targets for vision, and also to act as distance markers to help train the guards in estimating distances to the surveyed field at night.

4-7-5 For large industrial and commercial premises, proper patrolling within the building may be of great importance to security. In some cases the security guard will try not to be seen from outside the premises, so that minimum or no lighting will be used within the building; in some cases, the guard will observe the interior of the building from an external patrol path, or he will need to move about the buildings and check various matters, so that a system of pilot lighting will assist his work (see 3-10-2).

4-7-6 Considerable use is made these days of visiting patrols. The occupier engages a security company which undertakes to visit the site an agreed number of times per night and check that all is well. This can be an economical solution to the guarding problem, but it is unlikely to be effective unless the following conditions are met:

- It is utterly useless for the patrol to arrive and to switch on the security lighting while inspecting the site, and then to switch off the lights again as they leave. This simply sends a signal to all villains in the vicinity, telling them when it is safe to break in! Instead, the security lighting should be kept on during all the hours of darkness.
- If there is more than one entrance to the site or building, the visiting patrol should enter and leave by all the entrances in random order.
- Not only should the patrol's visits be at random times, but the number of visits should be randomised too. If this is not done, it is simple for a watching villain to count the number of visits, and then break in after what he is sure will be the last visit of the night. Randomising is achieved by contracting for say 5 visits per night, but with say an extra 10 visits per month to be distributed randomly over the month.
- The itinerary of the patrol's time should be flexible. If the watching villains know that the patrol, having inspected site A, cannot possibly be back in less than say two hours (because they have to visit site B), then the villain has a secure period in which to break in. It can be very constructive

for the patrol occasionally to return within a short time after completing a visit, and make a further inspection. This can be achieved with the cooperation of the security company, who will deploy two or more different squads to visit several sites, with some randomising of routes, times and destinations. It is better if the cars are controlled by radio from the security company's control room, so that even the patrol personnel do not know where they will be at any future hour through the night – thus making collusion with criminals more difficult. Radio messages directing the patrols can be coded to keep the information confidential.

4-8 Use of cctv and radiocommunication systems

4-8-1 Supervision of long fence lines and large areas is often essayed with the use of closed-circuit television (cctv). Even with the most sophisticated system, cctv has its limitations, for no-one can continuously observe cctv monitor screens hour after hour and maintain a high standard of vigilance. In tests, it was found that a guard watching four screens which automatically monitored twelve cameras (each screen changing to the next of three scan settings at ten-second intervals) became confused and virtually hypnotised after about nine minutes and ceased to be effective. Watching a single screen, hour after hour, was also judged to be ineffective because the guard simply could not continue looking at the screen, or fell asleep. It is for this reason that automatic picture-change monitoring (automonitoring) has been introduced (see 4-8-2). The most successful applications of cctv seem to be in augmenting visual supervision, and for the security guard on watch to monitor the progress of his colleagues as they patrol around the site.

4-8-2 Cctv is used to extend the range of supervision possible by plain sight. Television systems can be adapted to provide alarms too, by the incorporation of 'picture change detector' function. In these, the camera is trained upon a static scene, and it is arranged that the computer controller scans the scene at, say, 1-second intervals, and compares successive pictures for signs of change, i.e. the presence of an intruder, or a door having been opened, etc.(automonitoring). Where change is detected, the alarm circuit is triggered, and thus it is not necessary for the security guards to observe continuously for signs of intruders on the monitor screens.

For certain high-security applications of a military nature there could be a case for using special types of cctv cameras which operate on ultraviolet light, or at very low light-levels or which detect by infra-red radiation; but none of these systems gives the benefits provided by continuous security lighting. After all, if the alarm circuit is triggered by the detection system during the dark hours, lighting will be needed to deal with the situation. If a conventional cctv system is used, designed to operate on an average vertical illuminance of around 5 to 10 lux (which is in the range of normal low-cost vidicon cameras), a quite ordinary system of security lighting will provide light for the cameras as well as providing all its usual benefits to security (see 2-4).

Supervision and control of entry   75

4-8-3 The use of video recorders is of considerable value in automatically recording transient conditions, so that the data can be examined again. Video recordings can be played back at high speed while searching for the required part, or can be slowed down or frozen to study one instant in time. Recordings made automatically during an incident may enable photographs of the intruders to be obtained, and – even under poor conditions of rendition – reasonably accurate general descriptions can be deduced. If the video element is arranged to operate only when there is change occurring in the picture, it self-edits out the long periods of no change. A clock circuit can be used to imprint the time on every frame. Such a recording could be accepted as documentary evidence in a prosecution.

4-8-4 Control of access to car parks, especially at office blocks, is necessary to give protection to the vehicles and persons using the car park. Often, parking space is provide at the side or rear of the building, out of sight of the receptionist or gate security hut (see Figure 4.3). An idea seen at an office block in Paris was to place several cctv monitor screens in the front windows of an office, showing views of the rear of the building and the car park. Thus, every passing gendarme could see what was happening at the rear of the premises.

Figure 4.3 Vehicle movements on this industrial estate are monitored by cctv cameras. (Photo: Photo-Scan Ltd, Dolphin Estate, Windmill Road, Sunbury on Thames, Middx TW16 7HG)

76 Supervision and control of entry

4-8-5 The recent technical explosion in the telecommunications field will be of considerable value to security workers. It is now possible for a security guard to carry a small lightweight radiotelephone unit in his pocket which will not only keep him in touch with his supervisor and colleagues, but will enable him to phone out directly to the police or security agency for aid from wherever he is on the site, or from his car. Important numbers can be entered and will be automatically dialled on pressing two buttons only – a convenient procedure in the dark. There will be some situations where the guard wishes to keep silence, and it will be necessary for him to switch his radiotelephone off to ensure that his position is not signalled by bleeping.

4-9 Protection of security personnel

4-9-1 Risks of a guard being assaulted or injured

In the performance of their normal duties, security personnel are rarely placed at risk of being of being assaulted, but there will doubtless be those occasions when they find themselves in a serious confrontation with a violent or armed intruder. The risk will vary greatly from site to site, according to the support available, and the communications. How well the security guard deals with confrontation situations will largely depend upon his training. It will generally be better for him to rely on developing an air of calm authority to defuse the tension of angry confrontations than to practise his unarmed combat techniques.

The security guard may be put at risk of being injured in a personal accident when patrolling, particularly on construction sites and industrial sites where there is danger of tripping or falling or of striking some part of the body on sharp projections, e.g. reinforcement rods protruding from unfinished concrete work. The risks are always greater if the guard is working alone or without means of calling for assistance, and are especially great on unlit or insufficiently well-lit sites at night.

4-9-2 Reducing the risks

To minimise risks to security personnel, the following points should be considered:

• As a fundamental factor of the training and management of security personnel, it should be recognised that the protection of persons (including security staff) against risk of injury and threat to their persons is a prime objective of all security work. While it is hoped that security staff will show calmness and courage in stress situations, they should not be required to expose themselves unnecessarily to danger in the defence of property or in an attempt to restrain or detain a suspected person.
• Security personnel should be given the benefits of having safe working conditions – just as any other employee. Their patrol paths should be safe to move over at all times, i.e. made level and kept clear of obstructions, and fenced or railed-off from dangers of falls etc. The provision of adequate lighting is a prime requirement, and such may form part of a

system of security lighting (see 2-4, 4-7); also suitable fencing will improve security and make the guard's work safer (see 2-2).

Guards working in isolation are always at especial risk. On small sites where it would be uneconomic to employ more than one guard, arrangements should be made to give him the advantages of being able to call help quickly in an emergency. The following points should be considered:

• The guard could be provided with a personal radio or cordless telephone to carry with him at all times so that he can keep in touch with a reporting base or call the police when needed.

• A system of phoning-in at set times has some advantage, but only if the arrangement can be kept completely secret from outsiders.

• Where a guard is working in isolation in a potentially dangerous situation, e.g. on a construction site, he may be equipped with an automatic alarm device, i.e. one which sounds off (or dials-in and transmits a pre-recorded message to the police etc,) automatically if he should fail to re-set the system at predetermined times or after the elapsing of a certain number of minutes following setting. The cancel-alarm function can be performed by phoning-in from the security centre to a liaison point, or by use of a personal radio or cordless telephone, or by key-switches installed at selected points along the patrol path.

4-10 Keys and locks

4-10-1 It is useful to distinguish between the management function of key discipline (i.e. who should have access to what), and the security function (i.e. ensuring there is no laxity or abuse, and that keys are properly recorded and held only by authorised persons). There may also be a separate engineering function (i.e. ensuring the right type of lock is specified for each application, and that locks are checked, lubricated and replaced when required).

4-10-2 The compilation and upkeep of a clear and accurate Key Register is an essential. A record should be kept of the types and serial numbers of locks fixed to every door etc. This will be facilitated if a master plan of the premises is marked-up to give a unique number to every door. The register will show records of the door numbers, and the lock numbers/key numbers for each, with the names of the current keyholders and the dates of issue of keys. The number of keys held per lock may be anything from two upwards. The quantity of keys to be carried by persons or issued for use on demand can be greatly reduced if a careful study is made of systems of mastering and differing, upon which the advice of a first-class locksmith will be invaluable. It will be a helpful feature if the keys in sets of identical keys are engraved with individual serial numbers in addition to any identifying number; this will enable the keyholder to be identified if a key should be found. Advantage should be taken of key registration schemes offered by reputable lockmakers; these ensure that the lockmaker will not make or issue any further keys for that lock without written instructions from the registered owner of the lock.

The key register may be kept in a loose-leaf book or in card-index form, permitting easy updating and extension of the system. The key register must be securely stored in a strong fire-resistant cabinet or safe which is in responsible care at all times. Alternatively, the key register could be kept on computer disk (on a stand-alone computer which cannot be accessed remotely and with suitable precautions to prevent unauthorised local access.) A complete set of spare duplicate keys should be kept in another safe elsewhere, preferably with a copy of the key register – this being readily facilitated if the register is on disk. The fewer the number of keys on permanent issue the better; members of staff should not be issued with keys to be retained by them unless they are likely to make frequent use of them. It is often better to issue keys from the security office when they are required, and for them to be returned and signed in after use.

4-10-3 When a key is reported lost, immediate action must be taken to negate the value of that key should it be found or be in the hands of a dishonest person. This may involve immediately fitting an additional or alternative lock to the door to secure it. In the case of safes and strong-boxes locked by key locks only and without a combination lock, it may be simplest to remove the valuable contents and place them elsewhere. The Lost Key Routine which is devised by the security manager should be set out in plain language and a copy issued to each person who is an authorised keyholder.

4-10-4 For premises having a substantial number of locks, it will be a sound policy to engage the services of a qualified locksmith on an annual contract. He will then carry out routine inspection and lubrication of all locks, and be available on short-notice call-out at all times to come and open jammed locks, replace locks or supply additional keys.

4-10-5 Electronic locks

The new technology in locks is bound to make fundamental changes to security management techniques (see 6-5-5). The equipment and techniques are already available, and widespread adoption seems bound to follow as the systems get progressively cheaper. The essential features of electronic lock systems are:

- The 'key' is a plastic card or strip carrying coded information equivalent to the configurations of a metal key. This coded information is usually in the form of magnetic imprints which are not visible, or may be in the form of coded printed bars or punched holes. Without sophisticated equipment it is impossible to read the code or generate a duplicate key.
- The lock is operated by passing the key through a card-reader. This is a device which 'reads' the magnetic strip, bar-coded or punched-hole coded information on the key, and actuates the lock mechanism control circuit. The physical movement of the locking elements in the lock is brought about by electromagnets operated by the control circuit.
- The operation of a lock can be by remote means, i.e. the key card can be passed through a reader in one location to open or close a lock at another location.

Figure 4.4 (a) Recognition of a token used with the PAC 2000 controller which controls up to four doors. The token will trigger the controller if brought within 50 to 100 mm of the reader face. (b) Control unit with facility for the user to enter a PIN (personal identification number) for greater security. (Photos: Software Control Ltd, Green Lane, Romiley, Stockport, SK6 3JG)

80   Supervision and control of entry

- Locks can be 'parallelled', i.e. groups of locks can be locked or unlocked simultaneously by passing a single key card through a remote reader.
- There is an infinite number of possible 'differs', i.e. any number of locks could be created, each of which would have a unique key card which would operate that lock only.
- Locks can be 'mastered', i.e. two or more locks which have non-interchangeable keys can all be operated by one master key card which will open or close any lock in the mastered group.
- If magnetic strip cards are used, lock combinations can be swiftly changed, i.e. the combination can be revised by pressing buttons locally on the lock, or by passing a special setting-card through a remote reader which will change the combination. Immediately this is done, the card key previously used will not operate the lock. This is of particular interest in such situations as hotel management; a lost or unreturned room key can be negated in seconds by simply reprogramming the lock remotely from a central panel at the reception desk, the operator being able to issue new card-keys immediately. The lock with the new combination could still be operated by a master card key valid for a group of rooms, e.g. one floor, or for all the rooms in the system. Another form of electronic admission control uses a coded token which merely has to be brought close to the detector unit to register (see Figure 4.4).
- Safeguards can be built into systems of electronic locks; for example, attempts to pass the wrong key through the card-reading slot of a lock can be arranged to operate a remote alarm, or can be arranged to wipe off the coded information from a magnetic card. Disconnection of the control circuit from an electronic lock will prevent operation of the lock.

4-11 Control of movement within a building by zoning

4-11-1 The principle of applying fences in a scheme of zoning to restrict vehicle access is discussed in section 2-3-2. The idea can be extended to control of pedestrian movement within and between buildings. Essentially the idea is that most visitors and staff entering the defended premises do not need free and uncontrolled access to all parts of the buildings and land; access to sensitive areas should be restricted to those persons who actually need the access, thus excluding others who might have dishonest intentions. Zoning may be expected to produce the benefits of:

- reduced opportunities for planned crime;
- less temptation for persons to commit spur-of-the-moment crimes of opportunity;
- proper control can be exercised with fewer security personnel.

4-11-2 An example of the application of the zoning principle to control pedestrian movement within a large office complex may be given. Before internal zoning was introduced, various abuses had occurred at these premises, including unauthorised persons using the canteen and toilets, and there had been difficulties in excluding persons from certain departments handling accounts and sensitive information. In this instance the zones were three-dimensional, so that floor cut-out control circuits had to be introduced on the lifts to limit access to certain floors, and involved having an additional security post on an upper floor to intercept persons ascending by the stairs.

The problem of unauthorised access by the staircases cannot usually be solved by purely mechanical means, for fire-exit routes must be maintained free at all times (see 3-2). In such a case, if the illegal movement cannot be prevented by structural means or guarding, it will have to be detected by an alarm system or – if necessary – observed by the guards by the use of cctv (see 4-8).

Chapter 5
Protection of newly-built and reconstructed premises

5-1 Confidentiality at the design and planning stage

5-1-1 Even before the building exists, all matters concerning its security should be regarded as highly confidential. The persons involved in every aspect of design and planning of the project should be trustworthy, and all documentation and drawings should be kept securely. Every photocopy or print of drawings should be identified with a unique number, booked out and signed for, and accounted for during its use. Later issues of drawings should only be provided to the users on the return of earlier issues. At the end of its use, every document and drawing should be signed back in by a responsible person. Redundant documents and prints, and all draft documents, sketches and plans should be recorded as duly returned, and shredded before disposal.

5-1-2 Non-essential information should not be put on to drawings. For example, there is no need to identify components specifically when a reference number or descriptive code will suffice. Cabling drawings for a building should show the circuits to be wired in, but need not inform the electrician about the functions of the circuits nor the equipments they serve. It is sound practice to provide additional cores in multicore cable runs to make identification of circuits more difficult. Actual connections of coloured or numbered cores to security equipment should be carried out by one trusted engineer who will make a single record of the connections for future reference, and this document should be stored securely. If drawings must be sent by post, it would be wise to send the equipment specifications or key to the drawings separately, but it would be better practice not to send any confidential documentation or drawings by post, but to consign by courier. As it is impossible to prevent unauthorised copying of documents, every document should carry a clear statement prohibiting copying, perhaps in the form of a impression in red from a rubber stamp, or overprinting in red or some other colour, so that a photocopy will be instantly recognisable as such. The illegal copying of documents facilitates various kinds of crime (see 6-4-3 and 5-8-3).

5-2 Precautions when altering existing premises

When carrying out structural changes, and especially when these changes are intended to provide an improvement to the security plan, the preparations for and execution of the work could, in fact, temporarily or permanently increase risks. For example, one would have to be on guard for situations such as the following:

- The opening up of ducts, wall cavities etc. may enable unauthorised persons to see routes of telephone cables, alarm circuit cables, and of fibre optic channels associated with computer installations. Consider erecting temporary screens to prevent unauthorised persons seeing details of the work having security implications (see 5-8).
- Malicious persons or criminals might have opportunities to sabotage systems including planting of listening devices (see 5-8).
- Temporary openings made in internal and external walls may give the opportunity for criminal entry, as may temporary scaffolding, ladders etc. brought to the site in connection with the work.
- Efficiency of patrolling and guarding may be affected if temporary obstructions (e.g. builders' weather-sheets) or new structures etc. interfere with existing satisfactory lines of sight.
- Newly-installed H&V ducts or suspended ceilings may create new illegal routes or new hiding places for criminal persons or stolen goods.
- Structural changes may affect the soundproofing of internal walls or floors, which could create difficulties in maintaining secrecy or confidentiality in sensitive departments.
- During visits to the premises by unknown representatives of firms quoting for structural and other work – do not discuss security matters with them nor reveal any weaknesses. Issue information only on a 'need to know' basis. Only the appointed contractor need be given all the plans, and even these need not include wiring diagrams showing vital security circuits (see 5-1).
- When contractors' staff are on the premises, take precautions to limit their opportunities for crime, including instituting some temporary zoning if necessary (see 4-11).

5-3 Security during construction of new premises

When an organisation is expanding by constructing an extension or putting up new premises on an adjacent site, there may be considerable stress placed upon the directors and managers because of the diversion of their time and energies from their normal duties. If the security manager finds that he has insufficient time to liaise with the builders and to deal with the many matters relating to the ongoing construction project, he could consider taking on a temporary assistant manager or consultant to assist him during the period of difficulty. Here are some of the matters relating to security which will require management attention during the construction of new premises:

84   Protection of newly-built and reconstructed premises

- Set up a system of liaison with own senior management, the main contractor, the subcontractors, and the Crime Prevention Officer of the police. Do this at the outset of the project, and maintain contact to deal with matters before they become urgent problems.
- Instruct the main contractor regarding safe and suitable storage of stores (e.g. electrical and plumbing components) on the new site so that goods do not deteriorate by exposure to the elements, nor are subject to pilferage (see 5-6).
- Work out proper schedules of delivery times for the plant and machinery that is to be installed in the new building, and ensure that it does not arrive too early, i.e. before the premises can be made secure. Consider the use of security containers (see 5-6).
- Consider the need for a temporary system of alarms or cctv; would it be better to hire this equipment for the critical period rather than go to the expense of purchase?
- Can the organisation's own staff cope with the manpower requirements for guarding the site during the contruction phases? Would it be better to bring in a security company to attend to the patrolling until it is a convenient time to appoint additional permanent staff and integrate the security for the new premises with that of the existing buildings?
- Consider making the provision of security lighting a contract condition to be met by the main contractor from the outset of the project (see 2-4, 5-5).
- Lay down requirements for temporary fencing. Consider, could the permanent fencing for the site be erected early to provide protection of the site and its contents during the construction phases? (see 5-4)
- If the new premises are on a site adjoining the existing one, decide whether it is really necessary to allow free movement of persons and vehicles between the new and existing sites; would it be better for security to keep these as two separate zones? (see 2-3).

5-4 Fencing the site

5-4-1 Practice in the provision of fences for construction sites varies very much. At high-rise and city-centre sites a reasonable standard of fencing is usually adopted, but, all too commonly, the fencing is poor or no fencing at all is provided. Intrusion may lead to damaged plant and vandalism or theft of stores, and the costs arising from delay in completion can be heavy. It is unrealistic to rely on guarding only – even on a well-lighted site – to provide the necessary protection if the site is not adequately fenced. Every fence is climbable or capable of being breached in some way; but a sturdy fence prevents the opportunist from just walking in, and in combination with suitable lighting will give the watchman a reasonable chance of protecting the site from all but the most determined and nonchalant criminals.

5-4-2 Young persons and children who get on to building sites – by day or night – may be in grave danger. HMSO Guidance Note No.G37, 'Children on construction sites'[14], states that it is the duty of the site operator to

Protection of newly-built and reconstructed premises   85

provide adequate fencing to prevent those tragic accidents which still occur. The fence helps to keep the children out; the lighting will enable the watchman to spot them and deal with them if they do get in. In the UK, if any trespasser – child or adult – gets injured on a construction site, the site operator may be prosecuted under the Health & Safety at Work Etc Act, or the Occupiers' Liability Act[22]. The occupier might also be sued for unlimited damages under the Common Law by the injured party or by his relatives should the accident be fatal.

5-4-3 A strong wire-mesh fence, or even a well-installed chestnut paling fence, will usually give better protection than a more expensive corrugated-sheet-steel fence or a wooden planked fence – simply because the former are transparent and enable the guard within the site to see what is happening outside, and also enable passing police to keep an eye on the site at random times (see 2-2).

5-5 Power supplies, security lighting and emergency lighting

5-5-1 Mains electricity supplies to construction sites are no longer regarded as a luxury or just an aid to winter building. The requirement for an electricity supply to the site arises from the need to operate electrically-powered construction plant and hand-tools etc. as well as to provide light for safe movement and the efficient performance of tasks. If a supply is connected, for relatively small extra cost the site operators can gain the benefits of security lighting in enhanced protection against theft and vandalism at night. On guarded sites, the provision of security lighting improves the working conditions and safety of the security guard, as well as acting as a powerful deterrent to intrusion.

One should not assume that the provision of this lighting will solve all the security problems. Security lighting can only be fully effective when used in combination with good physical security (e.g. fences, security containers for valuable stores), and with efficient guarding (see 2-2-3). With an electricity supply, high-risk sites may be further protected by a perimeter alarm system and/or cctv (see 2-5).

5-5-2 Getting the power supply on to the site may present some practical difficulties, but these are usually easily overcome by the application of some fairly obvious arrangements (to comply with the Wiring Regulations[15] and The Construction Regulations[16]) to ensure that the intake and the distribution cables are positioned to be safe from damage.

On high-risk and strategic sites, there is often a need for lighting and electronic surveillance from Day One of the construction programme to frustrate preparations for future crime (see 5-8), yet constructors may overlook the fact that a mains power supply simply cannot be installed the day after they decide they need it. Though there may be power cables immediately outside the site – or even crossing it – the electricity board requires reasonable notice of the requirements. It is not bureaucracy, but the sheer logistics of the engineering to be carried out, that occasions delay in the board providing the supply. It is sound practice to give preliminary warning to the board that a supply will be required, even as early as the time of the application for planning consents. This can make for economies; for example, instead of having costly temporary cables installed and having these replaced by the permanent cables later, it may be possible to take the temporary supply for site operations through the permanent cables installed early for this purpose.

With electricity available, it will be possible to install general area floodlighting for security purposes early in the site's development, and then to adapt and develop this basic lighting system to suit the constantly changing configuration of the site and its changing risks. If the connection of a mains supply is delayed, and the security risks are high (and productivity and safety requirements are pressing), it may be necessary to use on-site generators until the mains supply is available (see 5-5-3).

5-5-3 On-site generation of power may be arranged according to the size of the site and the planned duration of use of the temporary supply. For example, on a major site there may be several generators, these being installed in temporary or permanent structures complete with switchboards and transformers; or use may be made of trailer-mounted generator sets.

The common need is for general area-lighting luminaires on masts; these may use high-intensity discharge lamps, e.g. high-pressure sodium (SON) lamps, which operate on a 240 volt supply. Additionally there may be a number of lower-mounted luminaires which will employ filament lamps, tungsten-halogen lamps or fluorescent lamps. For safety, these should be supplied at 110 volts via a double-wound transformer having its secondary winding centre-tapped to earth so that the maximum potential above earth on these lighting fittings and their cabling is 55 volts. Luminaires which are mounted within hand-reach height (e.g. lighting on walkways, stairways and in shafts), or which are relocated as required by site operatives, should be installed with BS 4343 industrial plugs, sockets and couplings[17] which may be safely connected and disconnected by competent persons other than trained electricians.

If a mains supply is not available, one way of getting lighting on to the site with the minimum delay and complication is buy or hire self-contained trailer-lights. These carry a diesel- or petrol-engine-driven generator, with a cluster of floodlight luminaires on a telescopic mast which may typically extend to 6 m, 8 m, or even 10 m height. The trailer-mounted generator set may incorporate a reduced-voltage output as described, or separate weatherproof unit transformers may be employed to supply low-mounted, portable lighting and tasklights at 110 volts.

Because site requirements change constantly – even daily – most lighting suppliers will decline to advise on security lighting for sites, nor design the lighting layout for temporary installations; but there are specialist hirers and suppliers of mobile lighting equipment and some electrical contractors who specialise in providing site lighting, and they will visit the site frequently to relocate luminaires and extend the installation as required.

5-5-4 On some sites it may be decided to provide generators for standby power to back-up the mains supply. These take a little time to come into operation, so batteries must be used to bridge any outage of supplies to essential services and key lights if a break cannot be tolerated. The general requirement to provide emergency lighting for escape under the UK Health & Safety at Work Etc Act[7] applies to construction sites.

5-5-5 A practical difficulty is that on a site where engine-driven generators are running, the noise level may be such that offence is caused to neighbouring properties. There are legal constraints which make it necessary to apply silencers and mufflers to keep the sound output to an acceptable level. At night, a generator will make it impossible for the security guard to hear small noises, so the provision of mains electricity would be an advantage in this respect.

5-6 Security containers

5-6-1 Security containers are strongly-made metal enclosures used on construction sites and within insecure buildings to contain valuable stores. Typically they range in size from that of a large toolbox up to sizes of 2.5 m × 2.5 m cross-section by multiples of 3 m up to 12 m long. For ease of transport they may be of 'skip' construction, or they may be constructed to comply with the Department of Transport's specification for transportable containers. Typically constructed from sheet steel of at least 0.8 mm thickness, these containers have a welded construction on a framework of stout steel angles and tees. Their joists and flooring are designed to withstand a loading of up to 900 kg/sq m. According to type, the doors may be on the side or end, and wide enough to facilitate the handling in and out of heavy items of plant or even to drive in forklift trucks and dumper trucks. Their construction is robust enough to withstand determined attack; for example, the doors are fitted with full-length hinges, and provided with deep frame flanges so that it is difficult to jemmy them open.

5-6-2 The need for secure on-site storage is acute on non-patrolled sites at night, and in rapid-build framed structures where the contractors may be installing technical services and plant on the lower floors before the upper floors are completed. Until the building can be made physically secure, valuable items can be stored in security containers in the building or close by. The need for such protection is clear; an ordinary joiner's toolkit these days can have a value exceeding a month's wages for the tradesman, apart from special electrically-operated tools. One cubic metre of small electrical installation parts can have a value of £1000/£2000; high-wattage electrical discharge lamps cost £40/£50 each, and may be delivered to site in hundreds. If these risks have been properly anticipated, one or more security containers on site will permit some flexibility in the times of delivery; stores can be put into the containers until required, where they will be protected against dirt, damage and damp as well as theft.

High-value items such as electric motors, switchgear, computers etc., can be given a **satisfactory standard of protection**. In some cases where costly plant is being purchased, it may be a good idea to specify that the supplier shall deliver it to site in a security container. In one case known to the author, some plant was delivered to site in a security container, but by the time the container was emptied, the structure had been built up and there was no way of removing the container from the building except by cutting it up.

Security containers may be purchased, leased or rented (some suppliers offer sale-plus-buy-back). The provision of security containers on site requires some anticipation on the part of the site management to ensure that the container capacity is available before the valuable stores arrive, and there must be space to park the containers.

When specifying security containers, consider the climatic conditions of use and the nature of the valuables to be stored. Containers may be insulated against solar heat, fitted with anti-condensation heaters, or provided with fans with vents which will permit ventilation without vulnerability to attack. Fire-protection cladding can be fitted, and containers can even be fitted with internal drenchers or sprinklers if the contents are liable to ignite spontaneously.

5-6-3 Security containers must themselves be protected against criminal attack. Strict key routines must be followed (see 4-10). The containers should be sited where they can be easily supervised from both inside and outside the site, and the location should be provided with security lighting (see 2-4, 5-5). It is not unknown for skilled thieves to use resources available on the site (such as power tools, acetylene-gas cutting plant or thermic lances) to break into a container; therefore such equipment should itself be locked in a container when work on the site ceases.

Cases have occurred where a site crane has been used by thieves to lift a security container over the fence on to a waiting lorry (see 5-7). Protection against this form of theft is given by suitable locks on the jack-leg supports (to prevent them being collapsed for transport), and by the use of some form of anchoring, e.g. the security container may be welded to steel members of the structure, or ground anchors may be used. A ground anchor is a large corkscrew device which is screwed into the ground, and then the container is placed over it so that the stem of the ground anchor protrudes in through the floor of the container where it can be securely fixed. With this form of anchoring, the thief can only move the container if he can first break into it to get at the fixing at the stem of the ground anchor.

Security containers may be fitted with an alarm. This may be a self-contained battery-powered audible alarm and independent of the site electrical system, or a remote alarm may be employed by a connecting cable to the container. This is an application for which a glass-fibre cable with pulse-coded light monitoring would be suitable (see 2-5), for cutting the cable will trigger the alarm circuit and the cut cannot be bridged.

Alarm systems for containers can be triggered acoustically or volumetrically, the latter being less likely to be falsely triggered by knocking. Passive infra-red detectors used within a container will detect an attempt to open the container by flame-cutting, sawing or drilling. The doors may be fitted with alarm switches, the circuit being fitted with the usual delay to permit authorised opening and cancellation of the alarm.

Protection of newly-built and reconstructed premises 89

While insurers generally approve the use of properly constructed and supervised security containers, they may impose their own conditions; for example, it could be an insurance requirement that the containers shall be sited within a 'citadel' (an inner fenced compound – see 2-3), and that there should be security lighting, an alarm system, cctv monitoring etc. If the site is not manned at night, they may require the containers to be anchored, and clearly visible from outside the site. They may also impose a limit on the value of goods which may be stored in one container.

5-7 Countering the risks to machinery, cranes and plant

If proper steps of protection are not taken, machinery, tools, cranes, mobile plant and vehicles on any site are likely to be the subjects of malicious damage or theft, or may be used in the furtherance of crime on that site or elsewhere. The precautions against the foregoing types of attack will of course include the familiar security measures of fencing, lighting and guarding, and possibly the use of alarms. Additionally, on closing down work on the site, take the following precautions:

- Immobilise cranes. Use chains and padlocks, or wheel-clamps. (Cranes may be used to move stolen goods. Many cases have occurred where goods have been lifted over fences using the occupants' own lifting gear.)
- Immobilise vehicles and plant. Fork-lift trucks, dumper trucks etc. should be chained and padlocked securely together in groups on shutting down. Chain single trucks to a stanchion. Additionally, mobile alarm systems can be used to protect a group of vehicles. (Vehicles have been stolen and driven through the walls of buildings to gain rapid access – see 6-2-3.)
- Store ladders within a locked building or security compound, or at least chain them securely. (Ladders are commonly taken from building sites to attack adjacent buildings.)
- Lock off hoists and lifts. (A motorised hoist on a site in the City of London was used by thieves to remove lead from the roof of the building being repaired.)
- Do not leave tools ready for the criminal. (Flame-cutting and thermic lance equipment has been stolen from unguarded sites to carry out safe-breaking nearby.)
- Thefts by the same building operatives and tradesmen who have worked on the construction project during the day are common; they may return in the night to steal tools, materials, electrical components, central heating and plumbing items, kitchen equipment etc. Take steps to protect partially completed buildings during the quiet hours. Finished floors should be properly secured to prevent easy access when the site is quiet. Consider using portable alarm units, battery powered, and set to detect selected acoustic frequencies (i.e. breaking glass, footsteps); a few of these compact units could be placed in locked buildings after the workforce has departed each evening, locating them randomly in different buildings and rooms each night.

5-8 Preventing the preparations for future crimes

5-8-1 At every stage of planning and constructing a new building or extension to existing premises, precautions must be taken to prevent the preparations for future crimes. For example, one must not provide opportunities for unauthorised persons to interfere with telephone cables and computer connections (see 5-2, 5-3). Given the opportunity, undercover experts could plant radio listening devices in the building structure or in its cable systems.

5-8-2 The modern thief or industrial saboteur may bring considerable technical skill to bear upon his work. For example, he may obtain a job on your site to sabotage it. This is what is believed to have happened in the setting out of turbines in a power station, the work being found to be faulty and requiring costly and time-consuming correction, the foreign engineer who was responsible for it having disappeared.

In another notable case, when the engineers began to pour the concrete to form a bank sub-basement bullion strong-room, the pressure of liquid concrete caused the complete safe-door and doorframe assembly (weighing several tonnes) to fall over, revealing that the hundreds of steel reinforcing rods which were supposed to be welded to the rear of the door frame had been cut through – doubtless in preparation for what could have been a spectactular and costly crime at some future time.

At another site, a last-minute check by engineers revealed that massive steel bars in a main soil pipe which were supposed to prevent entry via the building's sewer connections had been fixed so they could be easily removed from inside the sewer.

5-8-3 The security of drawings used during construction and modifications to premises is a matter of considerable importance (see 5-1). On large construction projects it is common for complete sets of sub-masters of the architect's and building-services engineer's drawings to be held (sizes typically from A3 up to A0) which may be printed by a dyeline copier on the site. Some builders and constructors use a simplified drafting system, with all drawings reduced to A3 or A4 size, and these are conveniently copied on an office-type photocopier.

On high-security and strategic projects, any means of copying drawings must be supervised as part of the security plan for the site, for drawings are valuable tools of crime. With detailed knowledge of constructions and electrical circuitry, criminals may later break into the premises with ease, and might be able to negate the intruder-detection systems. Information given on constructional drawings could also facilitate later acts of sabotage and bugging.

5-9 Security during the early days of occupancy

5-9-1 In the preparations for occupying the new or modified premises, the occupier should have made contact with the Fire Prevention Officer (FPO) of the Local Authority fire brigade in connection with fire escapes,

Protection of newly-built and reconstructed premises   91

emergency lighting and other matters pertaining to the issue of the Fire Certificate for the premises. At completion of the work, before occupation commences, it is a wise step to invite the FPO back for a brief inspection of what has been done, to gain the benefit of any advice he may have to give outside of official requirements.

Just before occupying would also be a good time to ask your insurer's representative to pay you a visit during final testing of all fire alarms, escape means, intruder alarms, emergency lighting and stand-by power generation plant. Similarly, if it can be arranged, a final check over the arrangements by the Crime Prevention Officer of the police would be very sensible; this would confirm that all essential crime prevention hardware has been installed, and be an opportunity to run through with him the routines that are to be instituted.

5-9-2 If there are written standing orders for security matters and fire precautions, it will be wise to check if any revisions are needed to take account of new patterns of fire alarms, door panic bars and other hardware that has been installed. During the proving of all the building-services systems and security systems (e.g. normal lighting, emergency lighting, standby power, heating or air-conditioning, lifts, telephones, intruder alarms, fire alarms) note any additions which may be required to the standing orders. Security exercises, familiarisation patrols, training and briefing of all concerned personnel should be carried out before occupation day. This will ensure that security personnel can handle all situations in their province from Day One, and know the locations of all essential equipment.

5-9-3 Mark-up a set of reduced-scale (preferably simplified) building plans with the locations of all essential equipment and plant. This is of benefit to the security manager, and will be invaluable during security exercises and training sessions. In event of any major incident, plans of the site may be needed by the Fire Officer handling a conflagration, or by police dealing with an incident of any kind. On these plans, or on another set, mark up the designated numbers of the doors and locks throughout the premises (see 4-10).

5-9-4 When normal occupation of the new premises commences, remember that there may be special risks in the early days: there will be new faces to be learned; it may take time for small defects in systems and hardware to be discovered and corrected. Remember also that many outsiders (tradesmen, removal men etc) will have been on the premises recently, and they may have knowledge that could enable them to commit crimes under the new conditions and before security routines have become efficient. Outsiders may have been issued with keys too, so it would be good practice to have the locksmith change all vital locks as the last event before the building is commissioned and handed over to the care of the security staff.

Chapter 6
Security strategies for typical premises

Note: In this chapter, the security strategies for certain types of premises are described. It is possible that some cross-pollination of ideas might take place, in that strategies and precautions described in this chapter may be applied to parallel problems in other types of premises.

6-1 Offices

6-1-1 Multiple-occupancy office blocks

Multiple-occupancy office buildings are especially vulnerable to daytime walk-in thieves (see 4-4). There are special risks of multiple break-ins following a single penetration because of the construction of modern office blocks, with ceiling cavities extending over whole floors (see 3-7). In many blocks, businesses are separated one from another by flimsy demountable partitions which give scant protection against intrusion.

This type of premises may tend to become less secure with time because of leases passing on to unknown new tenants, and uncontrolled sub-letting. It becomes impossible to control keyholding for external door locks. If the tenants in multiple-occupancy office blocks would share the expense, they could appoint a security manager for the whole building in the same way that some groups of tenants share the cost of a receptionist, switchboard and cleaning service.

6-1-2 Typical security risks at office premises

Office premises may be vulnerable when neither adequately staffed nor securely locked, e.g. early morning, if cleaners open the premises and then leave doors unlocked and rooms unsupervised while working elswhere in the building. Offices have slightly lower risk of being broken into during the silent hours than some other kinds of premises, but the cost of fitting good locks, alarms, safes etc. is clearly justified. Risks:

- Thefts of equipment. (Risk of theft of easily-portable items of high value, such as small computers, computer terminals and VDUs, electronic typewriters, facsimile machines, photocopiers, calculators and telephone-answering machines. A typical one-room professional office today can

contain £30,000 to £40,000-worth (1987 prices) of such attractive items, the whole of which could easily be transported away by the thief in the boot of a car – see 4-4-2).
• Thefts/destruction of confidential information. (A break-in might result in only minimal lost due to theft, but could totally disorganise a company by destruction of its records or computer data. Many companies go into dissolution after a destructive intrusion into their offices. It is difficult to obtain full insurance compensation for loss of data which, being intangible, has no intrinsic value.)
• Malicious damage by intruders. (Insurance can compensate for loss of profits, but cannot amend for the loss of time expended by executives and other staff in the chaos that follows a break-in accompanied by the now-common acts of vandalism, destruction and defilement of the premises.)
• Assaults upon females; other threatening or violent situations. (These risks are highest in multiple-occupancy blocks with direct access from the street to internal passageways, stairs and lifts without a front hall reception control – see 4-4-1.)
• Thefts of cash. (In some offices, even though cash is handled for wages etc., there is often an attitude of 'It could never happen here!' which denies the evidence of crime statistics. Offices should be equipped to handle cash as are banks, building society offices etc, with public access restricted to the outer side of the counter and the security screen – see 4-4-4, 4-4-5.)
• Thefts from staff (see 4-4-3).
• Fire. (Fires in office premises occurring at night may be caused by the carelessness or deliberate acts of intruders; but, it is not uncommon for staff to deliberately set fire to the office to destroy evidence of fraud or shortages of cash.)

6-2 Warehouses, storage buildings, industrial buildings

6-2-1 Vulnerability

With the vast modern markets for goods having high value per unit volume such as TVs, videos and tapes, computers and accessories, as well as traditional types of loot, e.g. spirits, tobacco, fashion clothing etc., there are now more opportunities than ever for large-scale thefts from distribution depots and from retail warehouses such as those selling home-improvement and similar goods (see 6-3). Many warehouses containing goods of immense value are sited on remote parts of industrial estates, or in quiet parts of big cities, and are usually unattended at night. Many modern warehouses are comparatively flimsy structures (typically steel-framed buildings clad in easily-penetrated sheet materials) which barely keep out the weather, let alone a determined villain.

6-2-2 Entry by trick (see also 1-3)

A common pattern for breaking into commercial and industrial premises is for one member of the gang to obtain entry on some pretext during the day, and to conceal himself until after closing time. Having no pressure of time, he may be able to disconnect any alarm system from within the building before admitting his accomplices. Countermeasures to this form of entry-by-trick include:

- Regulation of entry, and head-counting in and out;
- Sweep-search of premises at close of business;
- Double setting-keys on alarm systems, with one setting-switch inside the building and one outside.
- Fence-off the vehicle park from vulnerable buildings, and provide security lighting. Containers and vehicles left overnight to be sealed or padlocked by the warehouse staff. Record the registration numbers of all visiting vehicles.

6-2-3 Reducing the losses due to break-ins

Criminals may use the occupier's own vehicles and appliances (or mobile cranes and mobile construction plant stolen from nearby construction sites – see 5-7) as an aid to penetration or to remove property from the premises. The following precautions may be taken:

- Immobilise cranes and lifting gear when the building is unattended. (On a Midlands site, unauthorised use of a crane enabled thieves to remove a safe from an adjacent office building and load it on to a vehicle – which was also stolen).
- Lock up or chain forklift trucks at night. (Thieves who broke into a warehouse in London used the occupier's own forklift trucks to move his goods out through the roof.)
- Secure ladders etc. from improper use. (Fire appliances, ladders etc. left in yards have been used to gain access to buildings through their unbarred upper windows. Unchained ladders are commonly stolen to attack adjacent buildings. Ladders and escapes which may be required for fire use may be stored in an unlocked shed suitably alarmed.)
- Do not leave the tools ready for the intruder. (The safe in a engineering works was cut open with the occupier's own thermic lance equipment.)

6-2-4 Violent break-ins

A building containing valuable merchandise (e.g. tobacco products) may have a structural shell which can be easily penetrated by driving a vehicle at – or through – the wall. This method has been used in a number of daring raids on tobacco warehouses, where the thieves have stolen a lorry load of goods and been off the premises within five minutes from the commencement of the attack and before there was response to the alarms. If the building structure offers poor physical resistance to criminal attack, it may be defended mainly by restricting access to it and by having a lighted and supervised fenced zone around it. Heavy-duty palisade fencing with a massive ground-beam or plinth will resist this sort of assault (see 2-2-7). Alternatively, protection may be achieved by constructing a high kerb outside the walls, or erecting bollards. Some tobacco warehouses have been constructed on plinths to prevent this form of attack.

6-2-5 Warehouse fires

Warehouse fires frequently occur during the silent hours. Spontaneous combustion in ordinary manufactured goods is rare, being usually associated with chemical stores. Fires of unknown origin are often attributed to 'electrical fault', usually without justification. Accidental fires, e.g. those caused by a carelessly-discarded cigarette, do occur; but only rarely does it happen that the first signs of smoke or flame are delayed much beyond an hour or so after the building was last occupied (unless someone has contrived this).

There are two main motivations for deliberate fires – fraud and malice. Contrary to popular opinion, occupiers rarely attempt the crime of arson, for insurance companies are expert at detecting such acts, and these days the financial situation of the occupier will be examined to see if there would be any likelihood of fraud. What is far more common is for a dishonest employee to set the premises on fire in order to conceal shortages of stock or cash, or to bring about a convenient destruction of false records and accounts. Precautions against deliberate fire-raising may include:

- Fit smoke detectors, and possibly sprinklers or drenchers.
- Install fencing to keep vehicles and persons clear of the building (see 2-2).
- Install security lighting (see 2-4) to generally reduce risks and reveal smoke earlier.
- Arrange for the premises to be regularly inspected by on-site or visiting patrols when not occupied.
- Prevent accumulations of rubbish inside the building or near its outside walls.
- Keep all accounts and fraud-vulnerable documentation and records in fire-resisting cabinets or safes, and site these in a low fire-risk area.

6-3 Retail premises

6-3-1 Employee theft

The opportunities for employee thefts of cash can be made very limited by the use of a modern till system, tagging of goods, computer stock control and a reasonable standard of supervision. However, without the technical controls, theft can be committed with the connivance of an accomplice customer, i.e. items can be deliberately rung up at very low prices.

6-3-2 Customer theft

Some stores will suffer a 'shrinkage' of as much as 4% of turnover, most of which is accounted for by customer theft (shoplifting). Electronic marking of goods, stores detectives, security mirrors and surveillance by cctv systems may be employed. Vigilance by the staff is a very important factor in containing this type of loss. Audacious thieves equip themselves with specially-adapted clothing having enormous poacher pockets, or bring special shopping bags which have means of concealing stolen items. It is not possible to fit electronic alarm tags to all kinds of goods.

A stratagem used to steal small valuable items such as jewellery is to hide the stolen item in the store, and recover it at a later time. The typical method is to embed the stolen article in chewing-gum, and hide it under the edge of the counter. An accomplice may create a diversion while the item is being taken and concealed. If the thief is challenged when the shortage is discovered, he or she will have no stolen goods upon them and may be allowed to leave the store. An accomplice then comes to collect the stolen item some days later. Sometimes a preliminary visit is made to put the chewing-gum in position; so if such is discovered, vigilance may enable the thieves to be caught when they return to perform the actual act of stealing.

6-3-3 Cash theft

Good cash management will reduce the risks of an attack upon tills or the cash-handling area (see 4-4-4, 4-4-5). A particularly vulnerable time is during any failure of the lighting, so there should be an emergency lighting luminaire adjacent to every till. Electrical tills cannot be rung open during a mains failure, so cashiers should be instructed to shut the till promptly if the lights should fail.

6-3-4 Burglary

It is a common arrangement for deliveries to be made to unattended stores out of business hours, the vanman having keys to admit him to the premises. Abuse of this system is rare, but an unusual form of burglary has taken place at a number of stores where employees have breached trust and parted with keys to thieves who literally loaded vans with stolen goods during evening sessions when they were apparently loading stock on to shelves. Police and public looking in through the windows would have regarded the scene as normal, the lights were on, and the persons visible were wearing stores overalls and looking businesslike. The countermeasures to this form of burglary will include:

- Fit a time-lock feature to the alarm system.
- Fit an electronic lock (see 4-10-5) into which the manager keys a special combination number known only to him and the vanman.
- Out of hours, illuminate the store to a low level with lamps of an unusual colour, say with green fluorescent tubes; display a notice in the window stating, 'Please notify the police immediately if you see any person inside this store when the green security lights are on. A reward will be given.'

6-4 Hospitals

6-4-1 The special risks

Because of the wide variety of security risks associated with the modern hospital, preparation of a comprehensive security plan is rather like planning the security strategy for a small town. Within the perimeter of a hospital complex there are opportunities for many types of crime if suitable defensive measures are not taken.

In recent years there has been a significant and regrettable change in public attitudes, and particularly amongst the criminal fraternity, for hospitals were formerly regarded as special premises from which no right thinking person would steal; and hospital staff were accorded a respect that made them almost inviolate to personal attack. Sadly, this is no longer the case. Some criminals now regard hospitals as fair game for any kind of crime. Emotionally disturbed and socially alienated persons carry out violent and destructive acts against hospital premises and its equipment. Every hospital now needs to take steps to prevent physical attacks against its personnel both on and off duty in the hospital environs.

It may be conjectured that this situation is worsened by two factors; firstly, the rise in drug abuse leads to hospitals becoming targets for the theft of almost any kind of drug; and secondly, in our modern and enlightened way of caring for the mentally deficient and the emotionally disturbed, we no longer incarcerate these unfortunates to protect the public from them and to protect them from themselves. Modern therapeutic and control measures for the care of these unstable and socially incompetent persons are designed to modify their behavioural patterns in a beneficial way by treating them with tranquillising or mood-changing drugs. Physical restraint is applied only in the most extreme cases or during violent phases. Unfortunately, while modern methods of treating these socially disruptive and unhappy people are largely successful and very humane, there is a small minority of these persons who, either because they have not taken their pills, or have taken an overdose, or have combined their prescribed therapy with unauthorised drugs or alcohol, behave in an unpredictable and often very destructive manner.

Every hospital has experience of having to handle these people, who may at times be unreasonable and violent. They tend to create problems wherever they are; but the hospital may be the only place they regard as familiar and safe, and it is to the hospital they return, time after time – often daily – and frequently create difficult problems for the staff.

6-4-2 Countermeasures to the risks to staff

Management of unstable and irrational patients, and control of their movements and actions, may be aided by the following:

- A pass-key system may be instituted, with every authorised member of the hospital staff carrying a key which gives them admission to parts of the hospital from which patients and visitors are normally excluded except when accompanied by a member of the staff. The means of admission could be an electronic card key (see 4-10-5). High-risk areas (e.g. stores, and rooms where drugs are dispensed or handled) should be protected from public access by at least two doors, at least one of which will be openable only with a key issued to specially-authorised persons.
- The personal communication system used by the medical staff could be extended or adapted to include an alarm signal, to be operated by any person carrying a radio-bleeper device. According to the degree of sophistication of the system, operation may simply signal that an alarm situation exists, or it may annunciate the name – or even the location – of the person operating the alarm.

- A concealed alarm-button should be provided in every area where staff may urgently need assistance to deal with actual or threatened violence.
- The architecture of the building may be adapted to give protection-in-depth, i.e. to ensure that there is no easy route for an offender to escape from the building without passing a security checkpoint where he may be stopped, or at least passing the windows of an adjacent department, so that he may be identified later.
- Preferably, members of the staff of either sex should not work alone or with a patient out of earshot of other staff. Their safety can be improved in some situations by replacing doors with screens, curtains or vertically-hung walk-through strip-blinds, providing a reasonable visibility barrier, without creating acoustic privacy. The method has its limitations and cannot always be used in consulting areas; though these days, many consultations take place in open-plan clinic/treatment areas with only the minimum use of screens, some acoustic privacy being achieved because of the size of the area and its ambient noise-level.

6-4-3 Thefts of drugs

Every hospital has patients who are under drug therapy, quite apart from those being treated in special clinics for treatment of drug abuse. Many of the assaults upon hospital staff are by patients who have been refused drugs. Looking for drugs to steal is a common motivation for breaking-in, and for entry by walk-in thieves.

It is not suggested that the general standards of care in prescribing and dispensing of drugs are unsatisfactory; but it is recognised that there are frequent attempts to obtain prescription materials illegally from hospitals and medical practices.

It would be unrealistic to believe that hospital staff – including auxiliaries and lay staff – are not sometimes concerned in the theft of medical materials; most commonly, thefts are of addictive drugs. In overseas countries, cases have been known of large quantities of antibiotics being stolen by staff for direction to terrorist organisations who cannot purchase such materials legally, though such circumstances would be unlikely to apply in the UK.

Cases occur in which thieves break into dispensaries and store rooms, and steal considerable quantities of material – including some for which the thieves could hardly have any use. Such indiscriminate thieving suggests that the culprits were not well informed on the materials, though in other cases only specific types of addictive drugs have been searched for and stolen.

An effective way of stealing can be by manipulation of documents, so the paperwork connected with all drug movements should be subject to searching scrutiny. There must be continuous audit to ensure that orders are not falsified to obtain additional supplies, and to check positively that all consignments are actually received, accounted for, and placed in secure storage.

With the availability of high-quality photocopiers, precautions against the falsification of documents should be taken. Some protection can be given against a photocopy being passed off as the real document by printing documents in two or more colours, but now there are office copiers which will reproduce colours faithfully. One way of identifying documents is to apply an authorising mark by impressing them with an embossing seal, or – better still – with a perforating stamp. This could prevent persons in hospital or health authority offices issuing fraudulent orders. The equipment to apply such identification marks to documents must be carefully protected against unauthorised use.

Theft of prescription forms has long been known as a way of obtaining drugs, and departmental routines for protection of pads of forms must be enforced. The addition of a serial-number and the application of a distinctive authorising mark by embossing or perforating would increase protection against misuse of forms. Fraudulent changes in prescription forms and similar document can be protected against by the use of special ball-point pens, the ink of which fluoresces under ultraviolet 'black' light to a different colour to that seen when the ink is exposed to normal white light.

It is well known for drug addicts to impersonate another patient in order to obtain extra supplies of drugs; addicts may create several identities for this purpose. In the absence of the issue of national identity cards in the UK (as is the practice in many other countries), the only controls that can be applied are certain security routines operated by the Department of Health & Social Security, and comment on these here would be unwise. However, hospital identity card systems are employed.

Drug storage cabinets should provide a high degree of physical protection without unnecessarily impeding normal access, i.e. having very strong doors which can be removed or folded away during the working day. The fitting of alarm circuits to individual drug cabinets is a practical measure, though strident audible alarms cannot be fitted to drug cabinets in areas where patients are treated or accommodated.

6-4-4 Theft of stores

Control of theft from hospital premises is made the more difficult by the size and complexity of the premises, the presence of many vehicular and pedestrian entrances, and the fact that many types of goods in large quantities are received by many departments. For example, a major teaching hospital has daily deliveries of drugs, medical supplies, furniture, bedding, food etc., which have to be routed to many receiving points such as the pharmaceutical stores, catering stores, mechanical and electrical services stores, furniture and equipment stores etc. initially; and from these to several dispensaries, main and subsidiary kitchens, local workshops, stores and laboratories in many specialist departments and wards spread over a complex of buildings.

The larger the premises, the greater are the opportunities for theft of stores after they have been officially received, and particularly when they are being moved from the point of receipt to the consuming department or ward. Expediency or shortage of staff may result in parcels, tote-bins or cartons being left lying around in corridors awaiting a porter to move them.

When a consignment arrives at the destination department, there may be no-one on duty to receive it because staff have gone to lunch, or possibly the department is closed for the day or for the weekend. The goods may then be left outside the department unsupervised to await the eventual arrival of responsible staff. It is not surprising that pilfering (by hospital staff, by walk-in thieves, and sometimes by patients and visitors) occurs on a massive scale.

The magnitude of the problem will be seen from the following list of major items which were stolen from a London teaching hospital during a six-month period (1987 values): a large carpet, value £900, which had been temporarily taken up during redecorating and stored in an empty office; a tote-box containing sterilised sutures and surgical accessories and dressings, value about £3800, which probably was mistaken by the thief for another consignment containing hypodermic needles; a large wicker basket containing sundry pharmaceutical supplies for a medical ward, contents value about £1000, the loss of which delayed the administration of a vital drug to a very ill patient; sundry tools, and a large quantity of new tins of paint, total value about £450, stolen from an empty ward being redecorated; two large bales of new curtains which had been made to measure (value about £1700), and a complete van-load of highly-infectious soiled bedlinen.

In the same period, intruders broke into a laboratory and interfered with the electric plugs, resulting in the switching-off of a freezer containing biological samples. This necessitated some patients having to submit to further biopsy procedures.

Every hospital can provide a similar history. Sadly, these crimes expose patients to danger in various ways, including the introduction of infections, damage to equipment, and loss of vital drugs and materials leading to delays in treatment – all matters more important than mere loss of valuable stores.

6-4-5 Countermeasures to opportunist hospital thefts

Countermeasures to combat opportunist thieving on hospital premises may include:

- Make it a condition of orders that deliveries may only be made to designated receiving points between stated hours on specified days.
- Provide a 'stores pound', i.e. a secure room in which arriving goods and goods in transit can be left safely under lock and key when the stores receiving officer or the keyholder of the receiving department is not available. Goods delivered to the hospital should be signed for 'unexamined', kept sealed, and placed immediately in the stores pound to await collection by the designated recipient. Apart from one key held by a responsible officer for emergency use, the only key normally available to give access to the stores pound should be in the charge of one person who is made responsible for the operation of the pound.
- The routines for wards and departments to raise requisitions for stores should be followed. When the stores arrive at the ward or department, the person in charge should accept full responsibility for them and ensure they are put to proper use, or placed in safe storage. If the receiving department cannot accept the goods, they should be sent at once to the stores pound.

6-4-6 Night security in hospitals

Although hospitals have at least some departments working through the night and there may be considerable numbers of staff on the premises, it should not be assumed that the high level of occupancy will provide protection against crime in the night. Typically, staff will leave a room unlocked and proceed to another part of the hospital, leaving the contents of the room vulnerable to theft. Because of the shift system and turnover of staff, a strange face in the corridor will not be noticed. Large parts of hospital complexes may be unoccupied at night and not secured from intrusion.

In one hospital, staff had to fetch special equipment from a room on an upper floor from time to time throughout the 24 hours. To leave free access to that one room meant having to leave open doors leading from the stairs and lift lobbies, so that there was uncontrolled access to the whole of the floor, some of the unlocked rooms containing delicate and costly equipment. The administrative action required in a case like this is to re-locate the equipment to which access is needed during the night, and place it at a geographically convenient point. Then the floor at which it was previously located may be made secure during the quiet hours.

6-4-7 Impersonation; walk-in thieves

Because of the size of hospital premises and their operation throughout the 24 hours, they are vulnerable to walk-in thieves and persons passing themselves off as members of staff (see 4-4).

6-4-8 Control of entry into hospital grounds

In the UK, few modern hospitals have their grounds enclosed by a secure fence, and most have free access into the grounds at all times. Whether this will continue to be the case in future will depend on the difficulties that are encountered in creating peaceful and safe environmental settings for hospitals. Sadly, there are many cases of staff being attacked in the grounds when walking between, say, the ward blocks and the nurses' home during the night. In the USA, some hospitals have armed guards patrolling the grounds. The grounds of some hospitals in the UK receive patrolling attention from the police, if only from the occasional Panda car driving round the main internal roads.

Without creating marked visual impact, and without the need for great expenditure, it would be possible to apply the principle of 'zoning' to many existing hospitals (see 2-3, 4-11). This is possible, even if there is no secure perimeter fence, and would consist of partitioning the grounds by internal fences to create a series of enclosed areas which can be closed-off by gates when required. With good planning, this will not impede normal use, but may make some approach routes longer by diverting them to provide better supervision. Direct access for operational movements of staff, ambulances, visitors and patients need not be impeded. After all, if the laundry block is normally unoccupied at night, only rarely will anyone need access to it at three in the morning, and thus this block could be included in a zone that is locked-off at night.

102   Security strategies for typical premises

A system of exterior lighting will greatly increase the night security (see 2-3), and could provide environmental lighting for safe movement along the roadways and in car-park areas. This need not cause any intrusion of light into the ward windows at night. The provision of night security guarding of the grounds and buildings would probably be justified by the reduction in losses by theft and damage, as well as giving valuable protection to personnel.

6-5 Hotels

6-5-1 Liability for guests' property

Security measures are instituted in hotels to protect the proprietors against loss arising from theft or damage of both hotel property and guests' property. A disclaimer of liability for loss or damage of guest's valuables unless placed in safe custody in the hotel safe is usually made, but may not be legally effective if the guest proves negligence by the hotel, e.g. failure to take reasonable precautions to prevent the loss or damage.

Valuables not kept in a safe or strong room are usually covered by exception clauses in the insurance policy; full insurance cover is usually only given if the goods have been placed in a secure place, a receipt issued, and a record kept. The snag is that an effective receipt can only be given if the goods are properly described and their value is known accurately. This can lead to fraud, e.g. a 'gold' watch deposited by a guest may not be of the value declared.

Some solutions:

- Commonly, guests deposit sealed envelopes or parcels the contents of which are not disclosed to the hotel. One way to deal with these is for the hotel to apply its own additional seal or to enclose the package in an outer sealed envelope, and state on the receipt and in the safe register, 'Package No.xxx; contents unknown; value declared by depositor to be £xxx'. With proper control of the keys and/or combination of the safe, this should limit the opportunities for fraud by guest, employee theft, or conspiracy with a guest to defraud.
- Another method of dealing with guests' valuables that is favoured by some continental hotels is to offer the use of one of a set of small 'strong-boxes' installed in the reception area in full view of the public space. The guest holds the key, and the hotel staff do not know what the strong-box contains. Means are provided for the guest to fit his own padlock additionally if he wishes.

6-5-2 Illegal occupancy of rooms

In large hotels spread over many floors, or having motel-type chalets in the grounds, the responsible manager may not visit every part of the establishment every day, and neglect of security routines, or abuse by the staff, can occur without his knowledge. To prevent unauthorised use of rooms (especially those not used during the quiet season), frequent inspection is necessary. Abuse may occur by a registered guest admitting one or more extra guests to his own room without the management knowing; or hotel staff may admit an unofficial guest to a room.

Security strategies for typical premises   103

6-5-3 Intrusion

Some large hotels suffer from nightly intrusion into the bedroom floors by undesirable persons such as prostitutes, pimps, drug-dealers etc. They may move around the corridors at night accosting guests or will use the house-phone in the chambermaid's closet to systematically phone the bedrooms to offer their services. Most guests are too embarrassed to complain to the management. Some hotels have a cupboard built into the thickness of the bedroom door; a guest wanting clothes laundered, cleaned, dried etc. places them on the rail in this cupboard, and a small external flag is displayed so that the goods can be collected during the night by staff who open an outer door inset into the room door. Modern hotels no longer fit these doors because of thefts of clothes and because intruders whisper through the ventilators in these doors and accost the room occupants.

The problem is to know who is a guest and who is an intruder. Intruders may be well-dressed, well-spoken and very plausible, yet have criminal reasons for being on the premises.

Some countermeasures:

- Identification by room key or electronic admission card (see 4-10-5 and 6-5-5) is possible, though this intrudes on the privacy of guests if they are frequently challenged.
- In addition to patrols by the hall-porter's staff, senior staff to make randomly-timed inspections of the entire premises. Such patrols are best carried out by two or more persons to ensure that a 'sweep' is made of each floor, so that intruders cannot remain undetected by simply strolling about the corridors in an innocent manner until the inspection is over.

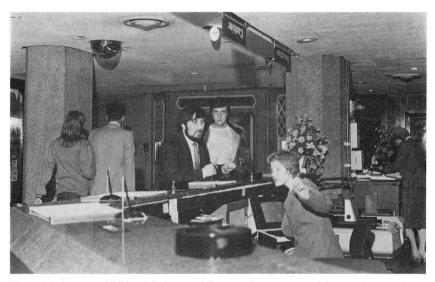

Figure 6.1 A recessed 'Videoglobe' cctv unit keeps a discreet eye on visitors at the reception desk of a well known London hotel. (Photo: Photo-Scan Ltd, Dolphin Estate, Windmill Road, Sunbury on Thames, Middx TW16 7HG)

104 Security strategies for typical premises

- As routine, every bedroom and reception room to be opened up during the morning cleaning and bedmaking routines – even if not occupied the previous night. Inspect every room, even during slack times.
- Fire exits to be fitted with alarm circuits to signal their being opened at any time, the alarm aranged to continue signalling until reset locally with a key. In suitable cases a 'glass bolt' type of lock may be used, or a 'Smash glass to obtain key' system employed, such provisions having to be approved by the insurers and the Fire Prevention Officer (see 3-2).

6-5-4 Thefts by persons impersonating staff

Thefts by staff from the bar and kitchen may occur if the management is not efficient; thefts from bedrooms by domestic staff soon result in the guilty person being discovered if the practice is repeated; but thefts from guests are sometimes carried out by persons who have got into the bedroom floors (as an intruder or as a guest), and then have put on cleaning-staff overalls which they have found a conveniently unlocked staff room or closet. Thus disguised, they are able to enter unoccupied rooms to steal (especially during the breakfast period). The stolen goods may be cached in an unoccupied room for later retrieval, or concealed in luggage if the thief is booked in as a guest.

Some countermeasures:

- In larger hotels, where all members of the cleaning and domestic staff may not be known personally to the shift manager and security staff, identity badges should be worn by all staff having access to bedroom areas.
- Staff rooms and closets where cleaners' overalls etc. are kept should not be left unlocked.
- Staff pass-keys should not be left hanging on a nail in the staff room or closet, nor in the domestic supervisor's office.
- The costs and practicalities of electronic card-keys should be examined (see 4-10-5 and 6-5-5).

6-5-5 Electronic card keys

The concept of electronic locks was introduced in section 4-10-5. These locks offer many advantages in the security management of hotels, including:

- The admission code for any room can be changed rapidly (possibly remotely from the reception desk), so that a lost or stolen card will not give admission to the room to which it applied.
- A current and valid room card key can be used to open the hotel front door after hours, or operate the car-park barrier.
- For staff use, 'differed' systems give the advantage that, for example, cards which will open store-rooms in the kitchen area will not open liquor stores in the bar area, and vice versa, yet both will open the staff entrance door and give access to other permitted areas.
- Cards can also serve the purposes of identification, and can carry the photograph and signature of the holder, as well as being electronically

coded to be used both as admission keys and for clocking in and out for supervision of timekeeping and wages computation.
• The issue of key cards to guests seems to be the answer to the problem of preventing persons strolling into the dining rooms of commercial hotels and getting a free breakfast by giving the waiter the number of a room they know – or hope – is unoccupied.
• The card system overcomes the problems regarding the handing-in and issuing of room keys. At a busy reception desk, staff cannot remember the face of every guest; so it is relatively easy for a cool thief to ask for a room key to gain admission to a room (a simple means of entry if the key-board can be seen from the public side of the reception desk).

6-5-6 Theft of baggage

Among his many duties, the hall porter has to supervise the handling-in and out of guests' luggage. A vulnerable time is when a party of guests – say, a coachload – is arriving or leaving, with perhaps upwards of 100 pieces of luggage being handled through the front hall. It is not unknown for an opportunist thief to put on a peaked cap and walk off boldly with two or three expensive-looking pieces of baggage during a coach loading or unloading operation.

At hotels which often receive guests in groups, it will probably be better if all baggage from a coach is brought into the hotel through a separate entrance directly into the baggage room, and not piled up in a vulnerable position on the steps or in the entrance hall.

6-5-7 Skipping

If a room is booked through a travel agent, or is booked by a local company for a visitor, the hotel can be reasonably certain of being paid. But when guests have not booked by letter, and have arrived without reservation, there is always a risk of them 'skipping'. Telephone reservations give some confidence, but only if from home or office and the reservations clerk can call back to verify the phone number.

The unknown guest who turns up with little or no luggage and no reservation presents the highest risk of skipping. Heavy spending in the restaurant and bar, with all charges to the room number, may give rise for anxiety. If the guest has little luggage, and – particularly – if he does not unpack but keeps the baggage in his room locked, the management must face a delicate decision: to ask for the account to be settled on a daily basis, or even to ask for a deposit. A bona fide guest should not be offended if the situation is explained to him, and will be willing (for example) to submit his charge card for checking at the time of arrival; but an unreliable one might bluster and fume, before perhaps tendering a doubtful cheque. Unfortunately, the day rate plus meals and drinks will always exceed the value guaranteed by a cheque guarantee card. Every hotel has experience of respectable-looking guests who disappear, leaving behind suitcases full of telephone directories or bricks.

Computer accounting brings the possibility of minimising losses due to dishonest guests. Hotels within a group or association can trade information about guests (including car numbers, even descriptions) with a few seconds of telephone transmission through the computer modem to other hotels in the network. Given the right program, a microcomputer can scan enormous lists of cheque numbers or cheque guarantee cards in seconds. Of course, this service is available centrally for credit cards, but it is not possible at present to check the validity of guarantee cards nor the ownership of cheque books, nor look for the guest's car registration number amongst a list of thousands of doubtful ones (which can be done in a few seconds with the right computer facility).

In a case of serious doubt, the police should be notified before the skipping occurs; certainly every case where a guest has skipped should be reported to the police who have means of collating such events with current crimes and their searches for wanted persons.

The guest's car registration number will be a valuable indicator to the police. Car registration numbers can be vital information in investigating the bona fides of guests, particularly if data exchange with other hotels is in operation. Because of its importance, the registration number should be recorded by a member of the hotel staff, or verified by inspection if the guest is asked to record the number on his registration card.

6-6 Security in outdoor environments

6-6-1 Open-air work and storage

Because of the high costs of owning and operating high-roofed wide-span industrial buildings, some industrial organisations which manufacture or process large objects, e.g. structural steelwork assemblies, pressure vessels, perform much of their work out of doors. Having no constraint of roof-height, they are able to use mechanical handling devices and cranes that would be impracticable to use inside a building, and in the open air they can employ engine-powered mobile machinery without the problems of air pollution from engine exhausts which would occur indoors.

Pollution and contamination are factors which lead other companies to do work out of doors which would be more hazardous for the operators in an enclosed space because of fumes or dusts, for example, the cleaning of chemical road-tanker vehicles, and steam-cleaning or paint-stripping operations on large industrial components. These open-air workplaces may be provided with fixed or mobile lifting machinery, and will have lighting to enable operation at all hours. Similar establishments are used as outdoor stores for products such as steel stocks, plastic piping and building materials; others are stores for palletised goods, each pallet being covered in a weatherproof shrink-wrapped plastic cover.

Because these outdoor workplaces lack the protection of solid walls and roofs, they may be very prone to thefts of tools and materials in the silent hours. Countermeasures may include:

- Provide good quality fencing (see 2-2). At points where the public can see into a work area or storage area, it may be advisable to put up sight-screens (lap-boarded or woven fencing inside the security fence, or perhaps canvas or sheet-plastic screens); but do not forget the protective

Security strategies for typical premises   107

value of casual supervision of the site by the public and passing police if they can see in (see 2-2-4).

• The lighting system used for the purposes of work might be adapted to serve as security lighting; the lighting level required for security purposes will be much lower than that required for working (see 2-4), and could be obtained by simply switching off unrequired lights. However, in order to obtain a uniform distribution of light without dense shadows, and in the quest for high reliability with low total cost, it could be more economical to employ an entirely separate system of lighting specially designed for security purposes. Note that, even though the installation is in the open air, an area might be declared to be a flame-hazardous zone requiring the lighting equipment and electrical installation to be of Flameproof or protected design.

6-6-2 Lorry parks

It is observed that in lorry parks which are well lit, the drivers tend to park the heavy goods vehicles much closer together than they do on dark sites. Because it is essential to leave clear lanes for the swift passage of fire appliances in emergency, it will be wise to mark out the whole area with white or yellow painted lines to guide the parking of vehicles. If the vehicles are stationed in straight lines, there will be clear avenues between them in both directions which will enable the whole site to be better supervised. Indeed, with chain-link fencing and a good lighting system, the whole area could be supervised from a patrol path outside the fence.

6-6-3 Docks, airports

The security problems at seaports and airports have much in common in that large numbers of persons and large volumes of freight and luggage have to be passed through the port with the exercise of a discretionary control of who and what passes in either direction. A common fault in the design and operation of security systems at ports is to make false assumptions regarding what would be the intentions of wrong-doers at the barrier; for example, while carefully scrutinising a mass of persons coming ashore, the officer might easily ignore a man threading his way with difficulty in the opposite direction through the crowd and passing through the checkpoint in the outward direction, particularly if the man was wearing gumboots and overalls and was carrying a bucket and a large broom!

At seaports, adequate care may be given to the screening of persons and cargos in efforts to detect movements of contraband, illegal drugs, explosives, arms etc., the efforts being concentrated at the times when ships arrive; but the sea is always there, and a much lower standard of vigilance may be accorded to the long, deserted sea beaches which border the controlled area of the port. Too much emphasis may be put on the passage channels, and too little on the dead areas wherein persons and objects might be hidden to await a more convenient time for crossing the customs line.

Persons of criminal intent will operate at all hours, and in all weathers. Three a.m., on a black night, with drenching, driving rain and a gale blowing, might be quite a good time (from the criminal's point of view) to be cutting the bottoms of chain-link fences round an airport service area or along a seldom-patrolled stretch of beach at a port, in preparation for illegally moving some goods.

6-6-4 Railway marshalling yards; container depots

The security problems at railway marshalling yards are made the more difficult by the fact that it is impossible to fence the area to be protected adequately. As it is impracticable to keep unauthorised persons out by physical barriers, the security strategy must concentrate upon early and more certain detection of the intruders. Major practical difficulties must be faced; the total area of sidings, vehicle parks, container parks etc. at a typical marshalling yard or road/rail container depot may be hundreds of hectares, so that complete supervision by patrols is impossible; and, because vehicles and rolling-stock are constantly moving in the area, electronic monitoring, even by automonitored cctv with 'picture-change-detection' (see 2-5-3, 4-8), cannot be used.

The best that can be done in such circumstances is to provide a high standard of security lighting over the whole area, and to maintain the best level of random patrolling by first-class security staff that can be afforded. At very large installations, an observation point at high level may be used, the observer being in close radio touch with the foot patrols and car patrols in the area.

6-7 Security at premises with especial risks

The general theme throughout this book has been to create good security by the provision of suitable physical protection and by proper guarding. Risks can be reduced by good engineering and by the use of alarm systems (see 2-5), and it is recognised that situations of particular risk may justify the use of more sophisticated means of electronic surveillance[18].

6-7-1 High theft-risk

In premises where high-value goods are handled and the risk of theft is high, control of loss will depend on instituting good physical systems, properly run key routines (see 4-10), thorough training of supervisors and managers (see 7-2), and the most careful selection of all staff in whom trust must be placed (see 7-10). Collect the data to enable differential analysis of crime-related information to be carried out (see 7-3-4). Give special attention to the following matters:

- Staff management:
   ○ Supervisors and managers to be on view – not hidden in offices or behind screens.
   ○ Where staff work in groups, change the composition of the groups frequently; re-allocate staff to alternative duties from time to time.

- Stock control:
  - Institute electronic stock control using hand-held computer devices and bar-code readers.
  - Carry out a partial stocktake each day, randomly selecting the sections to be checked.
- Zoning:
  - Apply zoning principle to vehicles entering premises to limit opportunities for goods to be transported away illegally (see 2-3, 2-8).
  - Apply the zoning principle to control of all visitors, ensuring that they cannot get access to any part of the premises not essential for the purposes of their visit (see 4-11 and Figure 4.2).
  - Apply the zoning principle to control of movements of staff, limiting their access only to those areas of the premises necessary for the performance of their duties.

6-7-2 High fire-risk

The prevention of fire is a major objective of security work (see 1-1-4). The task of security management may be considerably complicated by the presence of abnormal risk of fire as in factories concerned with munitions, fireworks, marine rockets and verey lights, and with the transport, storage, processing or manufacture of highly flammable substances, petrochemicals etc. Where the risk of fire or explosion is so great that the premises are defined as a hazardous area, special requirements apply relating to electrical installations and the performance of work – as regulated in the UK by the Health & Safety at Work Etc Act (HASAWA)[7]. An intruder may easily cause a fire or explosion out of ignorance or in defiance of the safety procedures (see 1-1-4), or deliberately out of malice, or to conceal evidence of a crime (see 6-2-4). The following observations may be helpful when considering security arrangements for such premises:

- If the premises are defined as a hazardous zone under HASAWA, persons should not be employed on security duties in and around them until they have been given thorough induction training in respect of the risks of fire or explosion, and have been trained in all necessary matters in relation to escape from the premises in emergency, fire-fighting and the rescue of persons.
- A problem identified by operators of such premises is in relation to smoking. It may be difficult to recruit non-smoking staff. It is even more difficult to be certain that a person who says he does not smoke is a genuine non-smoker. In these critical premises, the usual punishment for carrying matches etc. into a flame-hazard zone is instant dismissal – and possibly prosecution under HASAWA as well.
- A point often overlooked is that ordinary torches and handlamps are not safe to take into flame-hazard zones. Flameproof torches and handlamps are generally regarded as being too heavy to be carried on normal patrolling. Therefore it is essential that suitable permanently installed lighting is provided for the safety and efficiency of the security staff at night, this lighting using protected luminaires which are certified safe for the particular environment.

6-7-3 High terrorist-risk

The risk of premises being subjected to attack by incendiary device or explosives set by political or religious extremists, and the risk of key personnel being targets for assassination or being taken hostage, is not limited only to the obvious strategic, military and political targets. Companies trading in animal products, or involved in any kind of international trade, whether importing or exporting, may find that even their most routine commercial operations may be interpreted as having political significance; and they may be quite unprepared to cope with the risks to which they suddenly find themselves exposed when they may unwittingly give offence to sensitive parties.

6-7-4 The general approach to the defence strategy is summarised thus:

- Step 1: identify the vulnerable points. A vulnerable point (VP) is anything or installation which will be (a) costly to replace, or (b) difficult to replace quickly, and which will cause loss or difficulties of some kind if it should be destroyed. Accept that it is impossible to protect everything all the time, and select one or more VPs which you are determined to protect (even though this means that some less important assets must be given a lower standard of protection).
- Step 2: devise a strategy for the defence of the VP, remembering the importance of time to the attackers (time to cross spaces, time to overcome fences) during which they will be subjected to risk of being seen, recognised or arrested (for which lighting, patrols, alarm systems and communications will be needed). The reaction time of defenders or police must be less than the possible escape time once the presence of the attacker is detected, and the action strength must be matched to the potential attacking strength.

6-7-5 Defences against kidnapping, hostage-taking

There are special security agencies who can provide detailed advice on this complex subject, and vulnerable persons may need to be protected continuously. In the UK, civil and commercial organisations cannot use armed guards, so protection has to be achieved by other means. The potential subject may be particularly at risk when travelling to and from the place of work, and may need to be provided with a specially-trained driver. When devising the defences for the area where trouble might be expected, remember the value of depth in defence, e.g. the concept of concentric fence lines, zones, and redoubts (see Chapter 2).

6-7-6 Defences against bomb attacks

Special advice may be obtained from your Crime Prevention Officer on how to reduce vulnerability to attack upon persons, buildings, vehicles and installations. It is important that standing orders are devised for the conduct of persons on receipt of a telephone bomb threat, and that rehearsals and training sessions on how to deal with this kind of emergency are carried out.

Security strategies for typical premises   111

Anyone receiving an unusual or unexpected package through the post or by messenger, should treat it with suspicion. Unusual weight, oily stains, a strange smell, the presence of any kind of wires, a sound of ticking – all such things should raise doubts. If they suspect it might be a bomb, they should *not* open it, and should *not* put it in a bucket of water (which could explode it). They should leave the suspect package where it is, evacuate the area and call the police.

6-8 Public buildings, places of entertainment, sports stadia

6-8-1 Entertainment premises and sports stadia to which the public are admitted in large numbers suffer the risk of vandalism and damage by unruly crowds. Video recording will be of value in identifying and bringing prosecutions against individuals. A particular risk is that of deliberate fire-raising, when it may not be possible for the fire brigade to get access to and deal with a fire because of the presence of a crowd. The concepts discussed elsewhere in this book may be applied, e.g. to limit movement within the premises by the application of zoning (see 2-3), to provide a safe redoubt for persons who might be mobbed or attacked by a crowd, and to deal properly with the matter of cash handling (see 4-4-4). Throughout all these functions, the provision of the best possible system of communications will always make the tasks easier and more effective.

6-8-2 The control of entry into the premises will be a constant preoccupation of those managing them, and the use of coded membership

Figure 6.2 The control room of the Queen Elizabeth II Conference Centre where all the surveillance equipment is controlled and monitored. (Photo: Henderson Security Electronics, Unit 4, Tannery Road Industrial Estate, Gomm Rd, High Wycombe, Bucks, HP13 7EQ)

cards or admission cards which operate the turnstiles (see 4-10) should be considered. It will be necessary to formulate countermeasures to persons being admitted at fire exits by their friends who have already entered (see 3-2), and to deal with the problem of persons who conceal themselves on the premises at the close of activities (see 1-3, 6-2-2).

6-8-3 The operators of all places to which the public are admitted have responsibility to ensure the safety of all persons. The twin objectives of safety (fire prevention, crowd control etc.) and security (prevention of illegal entry, control of behaviour of persons on the premises, prevention of vandalism, theft and fire) are both served by systems of surveillance. In some cases, sophisticated monitoring systems may be employed (see Figure 6.2), but these should always be backed up with adequate numbers of trained staff who exercise the direct control and supervision of persons on the premises.

Chapter 7
The management of building security

7-1 Who is responsible for what, and to whom?

7-1-1 One of the most serious defects of management in organisations is the situation where an individual is made responsible for a function, but is not given the managerial and administrative powers to carry out that function efficiently and without constant reference to others for permissions and approvals. A common form of this is where the individual is required to perform some specific function or task, but is not provided with the proper means nor the spending power to acquire the means. Such situations may arise from the hierarchical power-tree which exists in many large organisations. An example is where some keys have been lost, and it is an urgent necessity to replace the lock, yet the person responsible for security has no power to spend a few pounds to protect the property by buying a new lock, nor the authority to instruct the works maintenance department to fit the new lock.

7-1-2 The officer given overall responsibility for security and protection of the premises, its staff and its contents should have a budget for forecastable expenses and labour costs, and should also have a smaller contingency spending power to deal with urgent matters and the unexpected. As that officer cannot be present 24 hours a day, part of this spending authority should be delegated to duty officers who may need to make urgent expenditure in his absence.

7-1-3 The security executive or manager of the organisation (that is, the director or senior manager who answers directly to a director) should be given a clear briefing on the objectives of the security function in that organisation (see Chapter 1) and accorded suitable spending power. There may be a discretionary budget which he can draw upon in an emergency, though in major non-urgent matters he may consult with a superior. A suitable routine of reports (see 7-3) followed by post-event sanctioning of special expenditures and normal audit procedures should prevent mis-use of this power.

7-1-4 It might seem to be a democratic and enlightened idea to allow free exchange of information between persons in the management hierarchy, and no doubt this works excellently in other management situations. But, in a department or section of the organisation which exists to combat internal crime and corruption as well external attack, it would generally be wise to require that information on security topics passes up and down the power-tree through 'normal channels' only. It is not unknown for persons high in an organisation to give illegal or improper orders to security staff, either for their own dishonest gain, or – perhaps in a mood of misguided laxness – to permit the bending of rules so that company loss or other disadvantage may occur. As an example of this: an electrician who was working late to repair plant urgently needed for production was asked to move his car from its normal daytime parking space in the staff carpark (which was not supervised at night) and place it in the supervised carpark used by the nightshift. He said that parking there with the shift workers would imply a loss of status, and appealed to his manager. The latter instructed the security staff that the electrician's car could remain in the unsupervised carpark. By an unlucky chance for the electrician, his car was stopped by the police when he was on the way home, and it was discovered that he had in the boot of his car a quantity of electrical cable which he had stolen from the works to do a private job. Had his car been parked in accordance with the works rules, he would not have had the opportunity to take the stolen goods to his car unobserved by the security staff.

7-1-5 The security specification should contain a clear scheme for reporting (see 7-7). This may mean that security guards and others will be required to sign a log book at specified intervals, and that the security supervisor may have to submit written reports in an agreed form to his superior. The small amount of clerical work involved is well justified, and the documents will usually be short enough to be written out in longhand without need for a typist to see them. Emergencies and suspicious circumstances should always be dealt with at once, and no essential security task should be delayed or omitted merely because the reporting procedure needs to be performed. If there is to be any scope for discretion to be exercised by supervisors responsible for security, the limitations of the discretion should be defined in the security specification.

7-1-6 While it is the responsibility of senior management to ensure that the security functions are identified and properly performed, it is not necessary for senior personnel to be informed of small details which would require their absorbing much information upon which they are to take no action. If a fully-detailed report of considerable length must be prepared, it should be prefaced with a very short summary which can be quickly read and understood by busy senior personnel without their having to read the minutiae which are not their immediate concern (see also 7-7).

7-1-7 Although many companies are organised on the power-tree principle, with lines of management extending from the Chairman down to the lowest-status employee, in practice there is also much 'diffuse management'. For example, although an employee may answer to a particular management line for the performance of his normal work, he may be technically in the charge of the canteen manager as soon as he steps into the canteen; the company safety officer may impose rules on his conduct in the works, or when using power tools or mechanical handling aids; as soon as he gets into a company vehicle he comes under the transport manager's rules, and so forth. Similarly, every employee must come under the security staff in all matters relating to their area of responsibility, be it in using the correct gate for entry, clocking procedures, parking rules, and their conduct during any emergency such as fire or an accident or a breach of security.

Because an individual has a high status in the organisation, even being of management grade, it does not follow that he is the best person to take charge of a security incident. Efficient response to a situation is more likely to be achieved if all staff – irrespective of status – will follow the instructions of the security officers as soon as required to do so. If this rule is not made and enforced, there will soon be persons 'pulling rank' on the security staff, whose position will be made difficult, and who will be severely hampered in carrying out their proper duties. If only for this reason, security staff must answer through proper line management to the board of directors, and the board should depute suitable powers to the director or senior executive responsible for the security function.

The situation is analogous to that relating to the police and society. The warrant card carried by a Chief Constable of a force is identical with that carried by a constable, and the constable has identical powers to his Chief under the law in matters such entering premises, arresting without a warrant, etc. The only power higher rank gives in a police force is the power to command the lower ranks in that particular force. So it should be with the security staff in industrial and commercial organisations, and every security officer of the company must be accorded due respect in carrying out his duties, for he derives his powers by the mandate of the board of directors.

If a security officer stops a senior member of the organisation to question him, to see his pass or search his car, the company rules should require that senior person to submit to the proper requirements of the security guard, exactly as he would for a police constable in a public place. It must not be assumed that crime is committed only by persons of lowly rank in organisations.

7-1-8 Security staff in most industrial and commercial organisations enjoy excellent relations with other personnel, and can usually count on their complete and willing cooperation in all safety and security matters. Unfortunately, there are a few organisations in which the security personnel do not get the respect and cooperation that their duties and responsibilities merit. In those few companies some employees may look upon the security staff as a kind of inferior police force, and they may exhibit an attitude of non-cooperation and or even overt hostility to the security personnel. The roots of this sad state of affairs are doubtless sociological and political, and reflect an attitude common amongst a very small proportion of the general public.

In such circumstances, the security manager must work hard at creating a good image for his staff, and do everything possible to build goodwill and gain respect. The problem is at least partly due to the fact that security staff have duties to prevent crime by other personnel; they must prevent frauds of time at the clocking-in bays; they must check vehicles and persons for stolen goods. The security staff are in a kind of parental role, and may be seen as the agents – or even the spies – of the management.

Where such unfortunate conditions prevail, they may be accompanied by a high incidence of petty crime and minor breaches of rules by staff who have a contempt for authority. Symptomatic acts, such as deliberately dropping litter, interference with fire appliances, neglect of safety rules, and aggressive graffiti, indicate that the organisation has a serious morale problem requiring urgent action and intervention from the top. To get matters right there is a need to communicate with the whole workforce, for it is in the interest of all employees that security shall be good and that fire rules and other safety routines be properly followed. If the general standard of security is poor (with frequent incidents of theft or malicious damage attributed to the staff), all employees are likely to suffer in the long term. The goodwill and cooperation of all staff in security matters is vital, not only to contain crime, but in order for the organisation to be able to cope with some future unexpected crisis – a bomb threat, terrorist attack, kidnapping, fire or major accident.

Even a workforce comprising carefully-selected persons with a high proportion of white-collar and professional workers will contain a few 'bad eggs'. It must be assumed that in every group of employees there will be one or two who might defraud the employer of time or money, who might steal, or who might commit some other kind of crime such as pilfering from colleagues. One comes across persons holding strong reactionary views who may erroneously feel that they have a right to help themselves at the expense of others, or who will not readily help with the performance of safety and security routines for the benefit of the organisation and other individuals. It will be the unfortunate duty of security personnel to act in the best interests of the employing organisation, even if this means reporting the illegal and antisocial acts of this small minority of employees.

Although the relationship between most employers and their employees is regulated by negotiation and representation and for most of the time is usually fairly harmonious, trade disputes do occur which cause such strength of feeling that persons who are normally of an honest and law-abiding nature will commit acts of defiance against the employer. In extreme cases, these acts will extend to destruction or aggression. The security manager and his staff must not believe that they have any choice in these stressful situations of conflict. Ethically and legally their position is quite clear, and their duties and priorities within the terms of their employment must be:

1 To protect all persons from injury or threat.

2 To protect the employing organisation and its members from preventable loss.

7-2 Training the managers

7-2-1 There appears to be an impression in some organisations that giving someone the title 'security manager' miraculously endows him immediately with knowledge of all the esoteric doctrines of security practice. Because this is patently not true, there may be a danger period when the newly appointed manager (unless he has previously been well trained and is well experienced in this work) may find he has to make urgent decisions on matters of which he has little knowledge. Without training he will have access to confidential information of various kinds, but will not yet become aware of the ease with which his lack of expertise could expose the organisation to risk.

It is important that newly-appointed trustees of sensitive information, and those who are given managerial responsibility involving any aspect of the security of the organisation, should be warned of the dangers, and should be committed to training as soon as possible after appointment. Preferably the new security manager will be given special briefing before the granting of his powers, i.e. a period of working in parallel with the retiring manager would be very helpful. In the case of new posts, training may be needed from external sources.

The appointment of the new security manager to his responsibilities may occur at a time of high activity and stress; there may be cogent reasons why the new manager should take up his post without delay; but no new manager (even a fully-experienced person coming in from another post) should assume responsibility for security of the organisation until he has received in-depth briefing and adequate training. The reason for this stipulation is that he will need to know full details of ongoing situations; of suspicions about the honesty of individuals; and he must have a proper understanding of the strengths and weaknesses of his staff, and of the strengths and weaknesses of the security installation. One thing he must not be tempted to do: he must not allow the security staff subordinate to him to teach him his job.

7-2-2 The post of head of security is a favourite career target for a retired police officer. But security management is not policing, although it contains an element of this and uses some of the same expertise. Nor is security management 'pure management' which can be learned as an academic subject. It is the art of combining the highest levels of personnel management with cunning and insight into how criminal acts of all kinds might be committed; it will also require an ability to liaise with management and staff of all levels throughout the whole organisation. Further, the responsible security manager will need to attain in-depth knowledge of his premises, and of all security installations (alarms, lighting, fencing systems, locks, key routines, patrolling system, communications etc.) as well as how to handle all kinds of emergencies including fire, and threats of all kinds (bomb threats, hostage-taking, etc). His responsibilities may include such diverse items as setting up and supervising security routines (i.e. weighbridge and pass routines); carrying out assessments of the efficiency of all aspects of the physical security of the entire premises; personnel management and budgetary control; study of the technology of alarm and communication systems; and receiving reports and statements, and making careful decisions for action in respect of these. Only by the exercise of these and many other skills will he be able to achieve the level of security needed by the organisation. To achieve the necessary skills in all these matters, the security manager may need to undergo training at some external source, such as that organised by a trade association, university or police force (training that should be constantly reinforced by attendance at seminars and exhibitions), and by study of the literature of the subject.

7-2-3 Much of the work carried out by security staff is routine, and tends to become boring. Patrolling on a wet night is not an occupation likely to raise the spirits. Yet, somehow, security staff have got to maintain alertness and vigilance; they have got to be suspicious of everything and everyone, and not allow laxness in the smallest detail of their duties. An important function of the security manager is to train and constantly retrain all the security staff, striving to capture their imagination, stimulating them to conscientious and alert performance of their duties, day after day, night after night. No matter how sophisticated the alarms and other security equipment, it is the character and conscientiousness of the security staff that actually ensures the proper protection of the premises. Complacency and boredom are the greatest enemies of efficiency in the guarding of premises.

7-3 Monitoring security performance

7-3-1 Information for management

Because management may remain unaware of serious breaches of security over a long period (1-2-3, 1-2-4), it is necessary to rely on positive reporting rather than 'management by exception'. If the latter philosophy is followed, management action will not be triggered except by major security breaches which cannot be concealed from the senior levels of the organisation. If positive reporting is required from subordinates, managers may have to suffer the boredom of regularly reviewing Nil Reports; but, at least, there will be a record in which those responsible for day-to-day operation of the security measures will have put down exact reports of all incidents – no matter how trivial – so that their significance can be reviewed in the light of later events.

As an example of this, consider the case of a large office block in which the lighting in the basement car-park was repeatedly damaged. The damage was attributed to 'children' (though none had been seen), and the lamps had to be replaced and lighting fittings repaired several times over a period of weeks. Because these losses were 'minor', they were not reported to management. After another spate of damage, the lighting was left in a broken condition for several nights over a weekend, and during this time a services cupboard located in the basement car-park was broken into and the power supply to the alarm system was interrupted. An automatic 'no-volts' alarm occurred, and eventually the bells were silenced by a mechanic from the alarm company who was called out, but the interference with the wiring was so great that he was unable to re-set the system. Later during that night, the premises were entered via a ladder placed across window-sills from an adjacent building, and a major theft took place. The break-in was undetected until next day. It would seem that the repeated

The management of building security 119

interference with the car-park lighting was in fact a preliminary phase of the crime. Had it been reported and its possible significance seen, the crime might have been prevented.

The seeds of the crime just described in this example were sown at a very early stage in the planning of this building and its services. Proper liaison between the building services engineer, the architect and various security experts would have thrown up the point that vital electrical services were to be housed in a chamber accessible from an unsupervised basement car-park with unrestricted entry from the street. This weakness could have been eliminated by re-planning the details of cables, and relocating the cabinets.

7-3-2 Positive reporting

At such periods as seem appropriate, management should be able to review a record of every incident which might affect security, no matter how trivial. If security guards are required to enter all incidents in a logbook and sign, date and time each entry, this information can be relayed rapidly to the person in charge of security strategy. A blank entry is not enough. The guard must certify 'No reportable incidents', and thus carry the responsibility for the entry.

When there is an incident which is identifiable as having possible security connotations, or when there is some attempted or successful crime against the premises, a full report should be compiled without delay, and the staff should not be allowed to go off duty until this has been done to the management's satisfaction. If a crime or serious incident has occurred, it should be routine to report this to the police in writing, even if they were not asked to attend at the time of the incident. Copies of all statements and other documents should be made, certified as to authenticity, and filed safely for future reference.

At monthly, quarterly and annual intervals, the record of the security-related events at the premises should be reviewed personally by the director or senior manager responsible. Periodically recurring data, and mass data of any kind, might be tabled or graphed into a digestible form for review – a simple matter with the ready availability of suitable computer programs. Examples of such data are numbers of persons and vehicles entering and leaving, analysed by gate number, day, time etc; the number of security staff on duty at various times; all incidents involving keys lost (or found), loss of any record; unauthorised or unexplained absence of any member of the security staff. Of course, a record of all visitors will be kept, and this should include their times of entry and departure, and the type of vehicle and its registration number. Data of this kind can be sifted by computer to reveal the presence of unlikely relationships which have great diagnostic value when analysing the circumstances surrounding a crime.

7-3-3 Monitoring the security performance

In addition to the desk review of all security activity, it will be a sound policy for the director or senior manager concerned to initiate unannounced tests and exercises at random times. These are to test that every aspect of the system is capable of working (i.e. electrical and mechanical equipment is in order) and that the routines and procedures laid down for the performance of guarding duties and dealing with incidents have been properly learned by the staff. It also enables the standard of recording of incident data to be reviewed.

The author visited a large industrial works for a night inspection of the exterior lighting, and was invited into the gate-house to await the arrival of the night manager. During a fifteen-minute wait, several persons passed through the gate-house, leaving the factory area and entering the car-park and then returning. These movements were with the tacit consent of the security on duty, who allowed these men to pass through without going through the clocking channel and without recording the times of their exits and returns. The time-card system was of the central-recording type, which provides a weekly pay calculation as well as an attendance record. This sophisticated system was negated by the permissive action of the security staff. In a positive reporting system they would have had to record 'Yes' to a specific question such as 'Has any person passed into or out of these works other than through the clocking channels?', and state the name of the person, his times of passage, and the reason for suspending standing orders to permit his movement through the security screen.

The services of an outside security agency could be used to test the efficient working of the perimeter fence system and the system for managing visitors. Without previous training and briefing, there would very likely be objections from the security staff at this form of delegated supervision of their activities, but it is vital that such tests are carried out without giving prior warning to the security staff, for they – as much as the equipment – are under test. All such exercises should be recorded in detail, bearing in mind that a real breach of security could occur during the time of the exercise. A security agency could also provide a 'visitor' from time to time, both during normal office hours and out of hours, to observe the way in which the security staff handle the situation.

As an example of how departure from laid-down procedures can permit a crime to be performed, consider the case of a warehouse where structural alterations were being carried out. Shortly after the end of the single day-shift, a van arrived, and the driver said he was delivering components to the builders who would want the materials early next morning. He produced no paperwork, but the guard allowed the van to enter and park out of his sight while the driver and his mate were ostensibly unloading. The guard was alone, and considered that he could not leave his post. When the van left, the guard was not aware that three men who had been concealed in the van were now free inside the works. During the night the intruders carried out a successful major theft which was not discovered until they had left some time the following morning. In this case, the guard did not even record the details of the vehicle. An employee at the works had seen the three men leaving the van, but was not aware that a crime was in progress, and did not report the matter until he was questioned by the police next day. Such crimes using the 'Trojan Horse' trick cannot succeed unless there is negligence or non-performance of security duties by the guard.

7-3-4 Differential analysis of crime data

Monitoring of security performance has two main functions: (a) to test and assess the performance of the security routines and actions by the security staff, and (b) to reveal any tactical weakness in the system, or need for new or improved equipment or standing orders to staff. By study of the monitoring reports, assessment may be made of the effectiveness of the management of the security function, and of the budgetary control, too.

The method of performing differential analysis is as follows:

- Devise a filing system and reference system so that reports and statements (see 7-7) relating to any incident can be quickly located, and so that papers on similar subjects can be collated.
- Keep a 'personnel log', i.e. a detailed log of persons on duty, sick, on leave etc.
- Keep an 'occurrence log', containing a brief record of everything that could possibly have a bearing on the running of the business and the control of crime.
- Frequently review the 'personnel log', the 'occurrence log' and the filed reports and statements about crimes and losses. Compare the date, time, location and nature of each crime or loss with concurrent entries in the 'personnel log' and 'occurrence log' (it may be possible to use a computer collating program for this). The objective is to look for coincidences, i.e. to determine if a certain sort of crime or loss occurs with greater frequency when certain persons are on duty (or are not on duty), or when there are concurrent external or internal events which might have some connection or which might create an opportunity for illegal acts.

Meticulous monitoring and analysis may demonstrate some surprising consistencies in the relationships of apparently casual events, and may assist in discovering how crimes may have been committed, and will be invaluable in helping to plan better precautions and controls.

7-4 Liaison with the police and security contractors

7-4-1 In the routine management of the security of premises, the manager would be wise to maintain a continuous contact with the local police, and especially with the Crime Prevention Officer (CPO) of his local force. The security manager should hold periodical meetings to review matters pertaining to his areas of responsibility (especially in the case of major premises) and it may be possible for the CPO to attend some of the meetings to advise. This is valuable when changes to the premises or the risks are taking place.

7-4-2 In many parts of the UK there are Crime Prevention Groups consisting of representatives of local commercial and industrial concerns who meet with representatives of the police regularly. Other bodies such as the Local Authority, local hospital administrators, Youth Service and education department may also send representatives. At such meetings the CPO may provide updates of information regarding local crime trends. Such a forum is excellent for the cross-pollination of ideas between security managers in the district, giving them an opportunity to study the patterns of crime and to set up collaboration in crime prevention matters. Broadly speaking, Crime Prevention Groups are to industry what local Neighbourhood Watch groups are to occupiers in residential areas, and the common experience is that where such groups operate the incidence of crime tends to fall. At meetings of local groups, there may be opportunities to set up joint schemes for shared-cost systems of fencing, lighting and patrolling of industrial parks and trading estates (see 2-7).

At meetings of local Security Groups, talks may be given by the CPO or other officers of the police, and sometimes contractors and suppliers from the security industry provide speakers, so giving security managers a chance to update themselves on systems and hardware.

Continuing contacts of this kind enable the security manager to develop a broader understanding of the work of the police. The publications of the Home Office Research Unit enable the security manager to gain some appreciation of the research work undertaken within the Home Office, these being issued to assist in the exercise of its administrative functions and for the information of the judicature and the general public[19].

7-5 Liaison with the insurers

7-5-1 Always bearing in mind that the main objectives of the security manager are the protection of life and the limiting of losses (1-1), it will be advantageous to the organisation if he can manage the security function with proper control of costs. If he performs his functions well, the result may be containment – or even reduction – of the costs of insurance.

7-5-2 Economy in the cost of insurance may be achieved by consideration of the following points:
- What additional or alternative measures of physical security could be introduced that would so reduce risks of loss as to merit a lower insurance premium? If capital cost is expended on physical hardware, or revenue is committed to increased manning costs etc., what would be the payback period due to reduced premiums (see 1-6-3)?
- After a suitable period of operation (say, annually), when the cost of the security operation and its effectiveness is reviewed, would it be a good idea to invite a representative of the insurers to be present?
- Are there any risks which have been so well controlled that the organisation should consider 'self-insuring', i.e. to carry the risk without insurance?
- In quoting their premium for the ensuing year, have the insurers taken into account all recent improvements in physical security, all changes in risk, and the loss record for the preceding period?

7-5-3 In many organisations the responsibility for obtaining insurances and paying the premiums rests with a director or a manager other than the security manager. In the budget for security, the cost of insurance is an important item; therefore it is suggested that the security manager should at least be consulted on all matters relating to insurance of the premises, and preferably that he be responsible for direct liaison with the insurers.

7-6 Management actions during an ongoing incident

7-6-1 In any emergency (e.g. fire, bomb-threat, hostage taken, armed robbery in progress, intruders on premises, etc.), all the persons involved or receiving information about the situation should know exactly what are their powers, and what they can do or authorise to be done before it is necessary to report up the line or to seek authority for further action. Only by much thorough training and rehearsal will it be possible for all concerned to achieve a swift and faultless response to an emergency situation. For example, telephone operators should have clear written instructions on how to handle every conceivable type of emergency, and periodic training sessions should be held to ensure that the instructions are clear and fully understood. It is essential that the policies are defined, and that the training is performed thoroughly, and that relief personnel and replacement staff are given induction training and briefing before their first day or night of actual duty.

7-6-2 In the case of a major emergency (i.e. one involving fire, threat to life, or risk of major loss), the security personnel and first-line supervisors should be able to take all essential actions necessary to limit the risk to life and risk of loss without need to refer to management. After they have taken these actions, they may need to make a brief verbal report to management, this to be followed up with suitable written reports when the emergency is over (see 7-7).

The best action for the security manager during an ongoing emergency is simply to be available to advise if required, and not to interfere unless things are really going wrong. The security manager cannot be on the premises 24 hours every day, nor perhaps even on call at all times; therefore he has to train his staff and, when the pressure is on, trust them.

7-7 Reports and statements

7-7-1 Because the exercise of responsibility can only be properly performed on the basis of correct information, accurate and prompt reporting of all incidents is vital. In cases of immediate and critical urgency (e.g. fire, bomb-threat, hostage taken, armed robbery in progress, intruders on premises, etc.), the responsible officer will probably receive a very brief verbal report initially (see 7-6); it is only when the incident is over that time will be found for preparing written reports.

7-7-2 If a report is prepared and submitted to a superior, it must be read and acted upon without delay, or – at least – a decision made on inspection of the report as to whether it requires urgent action or not. A report which lies in an in-tray for several days could contain vital information which, if acted upon, could protect the organisation from loss. An officer receiving any report concerning a matter of security should immediately decide if (a) the matter can be deferred for later attention, (b) the matter requires immediate action by the recipient, or (c) the matter must be referred to a higher level in the organisation.

7-7-3 Should an incident be likely to lead to a legal action, the preparation of accurate reports as soon as possible after the event will be regarded by the Court as tending to support strongly any verbal evidence given at the hearing. If any brief notes are made during or immediately after an incident (even scribbles, such a car registration number on the back of an envelope), such brief notes should be retained, and attached to and referred to in any formal statement made later. If a witness's statement is later copy-typed for clarity, the original statement should be retained and attached to the clear copy.

7-7-4 The preparation of written reports relating to security-breach incidents can be materially aided by providing security staff with a check-list or a reporting form designed to prompt the insertion of vital information. Police forces use such forms to report events such as vehicle accidents with damage only or with a personal injury, etc. Forms can be devised for use within the organisation for recording events which will occur from time to time, e.g. searches of persons and vehicles, operation or testing of alarms, defects in equipment, irregularities in key procedures, suggestions for improvement of security equipment or methods, etc.

7-7-5 Not everyone is gifted with the ability to write a good report. Many security guards and security supervisors who are competent in their jobs find difficulty in recording events for the information of the management or for use in later legal proceedings. If the person concerned cannot write a concise and accurate report, it could be helpful if he reported the facts verbally for someone else to write and for him to check and sign. In courts, witnesses sometimes repudiate statements they made as written down by a police officer, claiming that the words are not their own, or that the officer changed or omitted some facts. In training, police officers are taught how to take a statement, and instructed that their job is not to put in the statement what they think happened, but to write what the witness would write if he were able to do it himself. As in police procedure, a report or statement prepared at dictation should be read over to the witness, and the witness should have time to amend the report there and then, the amendments being initialled by the witness and the person writing down the report. If two or more people have to report on the same incident, they should not confer until their reports have been written and signed. Then, if there is a new matter which either thinks of on seeing the other's report, any additional material should be added at the end of the statement in a dated and timed postscript, and suitably initialled.

A useful way of preventing tampering with such documents is to make a photocopy straight away, mark the date and time on the copy and initial it, and 'post it' into a locked box with a slot, or hand it at once to a responsible person. In these days of rapid and cheap photocopying, there is no reason why any security staff or other persons who make statements should not have a photocopy to keep for their own reference.

7-7-6 The obligation to prepare a report on any incident should be written into the job specification of all security staff, for it is only by the study of reports from subordinates that senior managers and responsible officers can accurately and quickly appraise the details of events occurring in their province of responsibility.

7-8 Planning security management

7-8-1 The usual principles of good business management should ensure that the security plan is properly delineated, communicated to those who must carry it out, and fully understood by those who must administer and supervise the work. It may be a sound policy to create a security specification which enumerates fully all the objectives, describes the equipment, and outlines the methods and supervisory routines to followed in respect of security on the premises, including security for vehicles and outstation stores and depots.

7-8-2 All management communications to security personnel should be treated as confidential. It would be unwise to allow copies of the security specification to be available to all staff, but the information in it should be disclosed on a 'need to know' basis. Where it is necessary for written orders to be given, these should be succinct and communicated privately to those responsible for their execution, and such documents should quote the authority of the security plan document. Written instructions to security staff should not be pinned on the wall in the security room, nor displayed on notice-boards where others will see them.

7-8-3 It probably will lie outside the competence of the company's personnel to compile the security specification. This is an operation where an outside consultant might be employed to prepare the basic document. The consultant would work within board policy guidelines, and would take account of the advice of the Crime Prevention Officer of the local police, of technical specialists, and of the insurers. The content and structure of the security specification will depend upon the nature and size of the organisation and its premises and their location, and will probably include sections dealing with:

- Staffing levels for the security function, defining limits of responsibility (e.g. who can hire and fire), and giving titles to the holders of specific offices.
- Routines for inspection and care of fences, security lighting, alarm systems etc.
- Arrangements for regular inspection, maintenance and periodic updating of all aspects of the security system and features of the premises.
- Setting out clearly the links between the security staff management line and the board of directors.
- Defining the method to be adopted for reporting (see 7-1 & 7-7) and for budgetary control (see 7-1-1).

7-9 Confidentiality

7-9-1 All organisations have information that they would prefer to keep confidential. Thefts of information relating to the business of the occupier (industrial espionage) are very common, and – except in strategic premises where the Official Secrets Act and similar legislation apply – may be very difficult to detect and to prevent. Few organisations take positive steps to protect their secrets, not even to the extent of removing sensitive papers from desk-tops at the end of the day. There is a general view that 'It could never happen here', and that 'All of our staff are very trustworthy'. Industrial espionage is more often carried out by disloyal employees than by professional infiltrators, though doubtless some of these operate in the UK at the present time – especially in the high-tech fast-moving areas of development.

7-9-2 The kinds of information which are leaked or stolen range from plans to change prices, to information about research and the development of new products. On major contracts, the prices to be bid in competitive tender situations are particularly attractive to dishonest persons who may sell this information to a competitor and enable him to underbid. Seemingly innocuous information may have a black-market price, such as the salaries paid to key staff (useful information for a competitor or 'head-hunter' wanting to poach staff), and minor snippets of information about public-company activities when mergers, take-overs and new share issues are in the offing, for these are likely to affect share prices. A very damaging form of theft of information is stealing or copying costly computer data and programs (see 7-9-4).

7-9-3 Information may be leaked unknowingly by careless talk. Staff who are party to sensitive information should be particularly careful when talking to clients and representatives visiting the premises. If an office contains confidential records, or if data is displayed on wall-charts or is easily accessible through VDU terminals, it will be far safer to hold meetings with visitors in an interview room rather than allowing them to enter the executive offices or research areas.

7-9-4 The important and complex subject of computer line security is not dealt with here except in a general way in relation to line tapping and bugging (see 7-9-7). Theft of information from unprotected computers by those knowledgeable in their use is easy; it is equally easy to arrange for all confidential information to be locked-in by codes and passwords so it cannot be accessed by unauthorised persons. Computers can be programmed to shut-out all unauthorised enquirers from sensitive information, and arranged to report any attempt to obtain data without authorisation. Personal entry passwords should be so devised that it is impossible for a hacker to stumble on them by an intelligent guesswork; passwords (which can include figures as well as letters) should have no association with the user (not his house name, nor his pet name for his wife, for example). Passwords should be properly memorised by those who use them (and not written on a scrap of paper kept in a wallet or handbag, nor jotted conveniently on the blotter, nor pencilled on the wall by the VDU!).

7-9-5 Thefts of information do not usually deprive the owner of the data, but simply put him in the position of unwillingly (and usually unknowingly) sharing the data with other persons, known or unknown. Routes for such information include:

- accidental or inadvertent disclosure, perhaps involving a breach of trust but without deliberate criminal intent (see 7-10);
- deliberate theft or disclosure of confidential information by a trusted person;
- acquisition of information by visitors who are not properly supervised (see 4-4);
- theft of information in any form by intruders;
- sophisticated means of spying on the organisation may be used (electronic eavesdropping – see 7-9-7).

7-9-6 Unless vigilance and imagination are applied, confidential information may leak from the organisation by various physical means; for example, papers bearing confidential information (spare copies of letters, telexes) may be discarded without shredding; once-used carbon paper or carbon typewriter ribbons may be put in the garbage instead of being shredded or burnt.

Two examples may be quoted of technical information leaking from organisations. The first concerns a pharmaceutical company engaged in some important experiments. They sent some drums of spent chemicals to a disposal point for destruction, but the drums were surreptiously taken by a person connected with the company engaged in the destruction operation, who sold them to a competitor of the chemical company. The competitors analysed the contents of the drums and were able to gain some valuable information about the direction of the company's research. In another case, a company making diazo printing paper wanted to know the order in which its competitor applied several coatings to their paper. Unfortunately for their competitor, they discovered that he used the tails of coating-mill rolls for wrapping purposes, and these clearly showed the order of application of the coatings.

From such instances one can see that **waste materials consigned out of the premises may give valuable information to those who know how to look for it.** A policy of control and proper destruction of waste materials – even wrapping papers – may be important for maintaining confidentiality for certain manufacturing organisations.

7-9-7 The subject of electronic eavesdropping lies outside the scope of this book except that, in the management of premises security, no opportunity should be provided for technical spies to tap into telephone cables nor to plant electronic bugs in structures for purposes of eavesdropping. For premises in the high-tech rapid-development industries, and for military and strategic premises, vigilance during all operations of construction, repair etc. is essential, and it will be wise to restrict the information given to potential suppliers and installers (see 1-7-5). A time of risk from this sort of invasion is when additional security hardware is being installed by outside contractors (see 1-4). Particularly, great care must be taken to prevent bugging and tapping (and the deliberate weakening of structures etc.) during the construction phases of new or extended premises, and during alterations of existing premises (see Chapter 5).

7-10 Trust

7-10-1 Even if excellent measures of physical security exist, any kind of crime might be effected with the help or connivance of a person connected with the premises who breaks trust, or could become possible because of the neglect of duty or lack of vigilance by a trusted person. If there is a departure from consistent and meticulous performance of all security-related duties which allows a breach of security to occur, this may be attributed to negligence, carelessness, accident etc; but, if the breach of security occurs because duties have not been performed in compliance with both the letter and spirit of relevant security regulations, then this must cast doubt upon the integrity of those responsible. To achieve the proper performance of all security functions, many secrets may have to be kept. If information revealing potential or actual weakness in the security of the premises or in the regulations for its control passes to an unauthorised person, then the result can be as destructive as if the keys of the premises or the combination of the safe had been handed to a known criminal.

7-10-2 Security for premises is not achieved alone by engineering a sound building and installing surveillance systems; nor will a high standard of management and training of all concerned alone achieve the objective. The security of the premises rests ultimately on the integrity of those who perform the security functions and of those who oversee their work, who by their unswerving diligence and loyalty justify the trust placed in them.

References and further reading

1. CIBSE Guide, Section B20: *Security Engineering*; published by The Chartered Institution of Building Services Engineers, London.
2. BS 8220: – *Guide for security of buildings against crime*: Part 1: 1986 *Dwellings*; Part 2: 1987 *Offices & Shops*. (A further part, *Warehouses and distribution units*, is in preparation.)
3. BS 1722: – *Specification for fences*: Part 10: 1972 *Anti-intruder chain link fences*; Part 12: 1979 *Steel palisade fences*.
4. *The essentials of security lighting*, booklet reference EC 4131/383, The Electricity Council (available gratis from electricity boards in UK).
5. Lyons, Stanley L., *Exterior Lighting for Industry and Security*, ISBN 0-85334-879-0, Applied Science Publishers, 1980.
6. Lyons, Stanley L., chapter 'Security Lighting' in book *Developments in Lighting – 2*, edited by D. C. Pritchard, ISBN 0-85334-985-1, Applied Science Publishers Ltd., 1982.
7. Health & Safety at Work Etc Act, 1974.
8. BS 5266: – *Emergency lighting*. Part 1: 1975 *Code of practice for emergency lighting of premises other than cinemas and certain specified premises used for entertainment* (with amendments).
9. Technical Memorandum 12, *Emergency lighting*, Chartered Institution of Building Services Engineers, London, 1986.
10. ICEL 1001: 1978 *Industry Standard for The Construction and Performance of Battery-operated Emergency Lighting Equipment*, British Electrical & Allied Manufacturers Association and Lighting Industry Federation Ltd.
11. ICEL 1002: 1980 *Photometry of Battery-operated Emergency Lighting Equipment*, British Electrical & Allied Manufacturers Association and Lighting Industry Federation Ltd.
12. ICEL 1003: 1982 *Emergency Lighting Applications Guide*, British Electrical & Allied Manufacturers Association and Lighting Industry Federation Ltd.
13. Rayleigh, Lord. 'Night myopia', in *Collected Papers 1899-1920*, 6 vols, London: Cambridge University Press.
14. *Children on construction sites*. Guidance Note No. G37, Her Majesty's Stationery Office.
15. *Regulations for the Electrical Equipment of Buildings*, 15th Edn (also familiarly known as the 'Wiring Regulations'). Institution of Electrical Engineers, London, 1981.
16. The Construction (General Provisions) Regulations (1961)
17. BS 4343: 1968 *Industrial plugs, socket-outlet and couplers for a.c. and d.c. supplies*.
18. Walker, Philip, *Electronic Security Systems*, ISBN 0-408-01160-2, Butterworths, 1983.
19. Clarke, R.V.G. and Mayhew, P. *Designing out crime*. ISBN 0-11-340732-7, Her Majesty's Stationery Office, 1980.
20. *Code of Practice for Security Systems*, published by The Electrical Contractors' Association and The Electrical Contractors' Association of Scotland.
21. BS 4737: –, *Specification for intruder alarm systems in buildings*.
22. The Occupiers' Liability Act (1984).

Index

(The numbers in this index are paragraph numbers)

Access and exit, conflict of requirements, 3-2
Add-on security hardware, 1-4
Adjacent properties, 2-7
Airports, 6-6-3
Alarm situation, response to, 4-6
Alarm systems, 2-5
Altering existing premises, precautions, 5-2
Approaches to the premises, 2-1
Architect's involvement, 1-7-2, 1-7-3
Arresting criminals, 1-1-1

Baggage, theft of, 6-5-6
Bomb attack, 6-7-3
Bridges, 3-8
Bugging, 5-2, 5-8
Building services consultant, 1-7-3
Burglary, retail store, 6-3-4

Campaign for Nuclear Disarmament, 1-2-5
Card keys, 4-10-5, 6-5-5
Cash management, 4-4-4, 4-4-5
Cash theft, retail store, 6-3-3
Cctv, use of, 4-8
Ceiling voids, 3-7
Chain-link fences, 2-2-4, 2-2-6
Checking staff, 4-5
Checking vehicles, 4-3
Checking visitors, 4-4
Checklist for review of security, 1-8
Citadels, 2-3
Citizen's arrest, 1-1-1
Cladding of building, 3-3
Common Law responsibility, 1-1-3
Communication, 1-8
Confidentiality, 5-1, 7-9
Construction site, fencing, 5-4
Consultation on security strategy and specification, 1-7-3
Container depots, 6-6-4
Containers, security, 5-6
Contract arrangements, 1-7-5
Control of movement by zoning, 2-3, 4-11
Costs of security, 1-5
CPO, *see* Crime Prevention Officer

Cranes, on sites, 5-7
Crime data, differential analysis of, 6-7-1
Crime Prevention Officer, 1-4, 1-7-3

Degrees of protection, 1-1-1
Design stage, confidentiality at, 5-1
Differential analysis of crime data, 6-7-1
Difficulties of operation, 4-2
Docks, 6-6-3
Doors, 3-5
Drugs, theft of, 6-4-3

Early days of occupancy, security during, 5-9
Economics of security, 1-5
Electrical power supplies, 3-9
Electrical services, 3-9
Electronic card keys, 4-10-5, 6-5-5
Emergency lighting, 3-10
Entertainment, places of, 6-8
Entry, definition of, 1-1-2
Entry, methods, 1-3
Entry by trick, 1-3, 6-2-2
Especial risks, 6-7
Extended premises, 5-3
Exterior security lighting, 2-4

Failure of security, 1-2
Fencing, construction site, 5-4
Fencing and walling, 2-2
Fence-line layout, 2-3
Fence zone, 2-1-1
Fire, 1-1-4, 6-2-4, 6-7-2
Fire Prevention Officer, 1-7-3
Floors, 3-4
Foundations, 3-4
Future crimes, preventing preparations for, 5-8

Ground anchors, 5-6-3
Guard dogs, 1-1-3
Guests' property, liability for, 6-5-1

Hospitals, 6-4
Hostage taking, 6-7-3
Hotels, 6-5

131

Index

Impersonation, 6-4-7, 6-5-4
Information, theft of, 7-9-5
Infrequent risk, 1-4-3
Insurance, 1-5, 1-6-3, 1-7-3, 7-5
Interior walls, 3-7
Intrusion, into hotels, 6-5-3

Keys, custody of, 4-10
Keys, improper possession of, 1-2-1
Kidnapping, 6-7-3

Liaison with insurers, 7-5
Liaison with police & security contractors, 7-4
Lighting, emergency, 3-10, 5-5
Lighting, interior, 3-10
Lighting, pilot, 3-10
Lighting, security, external, 2-4, 5-5
Locks, electronic, 4-10-5, 6-5-5
Lorry parks, 6-6-2
Loss prevention, 1-6-3

Machinery on sites, 5-7
Management, planning security, 7-8
Management actions during an ongoing incident, 7-6
Managers, training of, 7-2
Man-traps, 1-1-3
Marshalling yards, railway, 6-6-4
Methods of entry, 1-3-2
Monitoring security performance, 7-3
Motivation for crime, 1-6-2
Multiple-occupancy offices, 6-2-1

New premises, 5-3
Night security in hospitals, 6-4-6

Objectives of security systems, 1-1
Offices, 6-1
Ongoing incident, management actions during, 7-6
Outdoor environments, security for, 6-6

Paling fences, 2-2-4, 2-2-5
Palisade fences, 2-2-4, 2-2-7
Parking of vehicles, 2-8
Partitioning, 3-7
Passing-off, 1-3-1
Patrolling, 2-6, 4-7
Patrols, visiting, 4-7-6
Pilot lighting, 3-10
Planning security management, 7-8
Plant on site, 5-7
Power supplies to site, 5-5
Precautions when altering premises, 5-2
Preparations for crime (preventing), 5-8
Protection of security personnel, 4-9
Public footpath, right-of-way, 1-1-2

Queen, HM The, 1-2-2
Quotations and tenders, 1-7-4, 1-7-5

'Rabbit runs', 1-2-3
Radiocommunication systems, 4-8
Railway marshalling yards, 6-6-4
Redoubts, 2-3
Reports and statements, 7-7
Response to an alarm situation, 4-6
Responsibility, for what, to whom, 7-1
Retail premises, 6-3
Reviewing security, 1-8
Risk, 1-6
Risks, premises with especial, 6-7

Safety of persons, 1-1-3
Searches, personal, 4-5
Searches, of vehicles, 4-5
Seasonal changes aiding illegal entry, 1-9
Secrets, loss of, 1-2-1
Security consultant, 1-7-3
Security containers, 5-6
Security lighting, 2-4
Shrinkage of stock, 1-2-1, 6-3-5
Skipping (from hotels), 6-5-7
Smoking and risk of fire, 1-1-4
Sports stadia, 6-8
Staff, checking of, 4-5
Standby electrical power, 3-9
Statements, 7-7
Storage buildings, 6-2
Structural enclosure, functions, 3-1
Surveillance, 2-6
Surveyed zone, 2-1

Television, closed-circuit, *see* Cctv
Terrorism, 6-7-3
Theft, of drugs, 6-4-3
Theft, retail customer, 6-3-2
Theft, retail employee, 6-3-1
Topography aiding illegal entry, 2-9
Trailerlights, 2-4
Training the managers, 7-2
Trespass, 1-1-2
Trick, entry by, 1-3
Trust, 7-10
Tunnels, 3-8

Vehicles, checking of, 4-3
Vehicles, control and parking of, 2-8
Violent breaking-in, 6-2-3
Visitors, checking of, 4-4
Voids, ceiling, 3-7

Walk-in crime, prevention of, 4-4, 6-4-7
Walkways, 3-8
Walls, interior, 3-7
Warehouse fires, 6-2-4
Warehouses, 6-2
Weather (aiding illegal entry), 2-9
Weld mesh fencing, 2-2-4
Windows, 3-6

Zoning, 2-3, 4-11